Zoophysiology   Volume 30

Zoophysiology

Werner Peters

Peritrophic Membranes

With 80 Figures

Springer-Verlag
Berlin Heidelberg New York
London Paris Tokyo
Hong Kong Barcelona
Budapest

Prof. Dr. WERNER PETERS
Institut für Zoologie
Universität Düsseldorf
Universitätsstraße 1
W-4000 Düsseldorf, FRG

ISBN 3-540-53635-3 Springer-Verlag Berlin Heidelberg New York
ISBN 0-387-53635-3 Springer-Verlag New York Berlin Heidelberg

Library of Congress Cataloging-in-Publication Data. Peters, W. (Werner), 1929 –
Peritrophic membranes / W. Peters. p.   cm. – (Zoophysiology; v. 30) Includes biblio-
graphical references (p.   ) and index. ISBN 3-540-53635-3 (Berlin). – ISBN 0-387-53635-3
(New York) 1. Peritrophic membranes. I. Title. II. Series. QL863.5.P47 1992   591.4'3 –
dc20   91-15512

Typesetting: International Typesetters Inc., Makati, Philippines
31/3145-5 4 3 2 1 0 – Printed on acid-free paper

Preface

The present review shows that peritrophic membranes occur widely in the animal kingdom, that they have many functions and that they deserve more attention than usual. It is unfortunate that many zoologists have never heard of peritrophic membranes or relate this term only with insects. Richards and Richards (1977) in their review of peritrophic membranes in insects complained about the vagueness and uncertainty in many aspects of these structures. Therefore, it seems to be time to give the first comprehensive review of our knowledge about peritrophic membranes in animals – and not only insects. The state of the art in this field requires a reliable source of data which can help to answer open questions, and which can facilitate search in the literature, in order to help those who want to solve problems with modern methods.

There is a bulk of information concerning morphological aspects, but there is a gap of information with respect to modern and functional aspects – or as A. G. Richards pointed out: "It is time for PM workers to become knowledgable in current methods of cell biology and to apply such methods" (Richards and Richards 1977).

WERNER PETERS

Acknowledgements

The author is very much obliged to all those who helped to complete this book: the editor of this series, Prof. Dr. Langer/Bochum, for the suggestion and constant encouragement, Dr. Czeschlik and Springer-Verlag for their patience, the Deutsche Forschungsgemeinschaft/Bad Godesberg for the support of investigations in a neglected field, Dr. Walldorf/ Düsseldorf for the line drawings, Mrs. Hanne Horn for her assistance in photography, and especially Mrs. Inge Latka for her skillful technical assistance in all fields of electron microscopy for many years. The author is moreover grateful to all those colleagues who kindly lent original photographs and unpublished results. I would also like to thank all my former students who worked on peritrophic membranes and contributed a large amount of information. Furthermore, I would like to thank Prof. Dr. Ulrich Zimmermann, Würzburg, for several years of stimulating cooperation. Much of this book was written during my sabbatical, and my thanks are due to all colleagues who took over my duties during that time.

WERNER PETERS

Contents

Chapter 1

Introduction

Membranous material, surrounding food and food residues in the midgut of insects, was observed already in the eighteenth century. Lyonet (1762) described it for the first time from caterpillars living in the wood of willows, probably from *Cossus cossus*. Ramdohr (1811) found "inner membranes", as he called them, in the midgut of *Hemerobius perla* (Planipennia), *Acanthocinus* (*Lamia*) *aedilis* (Coleoptera), *Malacosoma neustria* (Lepidoptera) and in *Vespula vulgaris* (Hymenoptera). During the nineteenth century such membranes were observed not only in numerous insects but also in other arthropods. Tulk (1843) found them in harvestmen (Opilionida), Darwin (1851/1854) in Cirripedia, Plateau (1876) in Chilopoda, Opilionida and Araneae, Gibson-Carmichael (1885) in a chilopod, and Schneider (1887) in various insects, in the water flea *Daphnia*, and in a chilopod and a diplopod. He claimed to have seen them also in snails of the genera *Limnaea, Helix* and *Limax*, but he gave no detailed description of these.

Nomenclature

In regard to how indiscriminately the name peritrophic membrane is used, it seems very urgent to start with some preliminary, clarifying remarks on nomenclature in this field. It will help to describe the different types of membranous secretions. The name *peritrophic membrane*, which means membranes surrounding the food, was introduced by Balbiani (1890):

"This is the name I propose to give to a sort of membranous sac which directly surrounds the food in the lumen of the intestine."

This definition can still be used as it is not only concise but also comprehensive. However, some remarks should be added on this occasion. The word "membrane" is misleading in that the peritrophic membrane has nothing to do with a plasma membrane, but is a more or less complicated secretion product. A structural element which occurs in nearly all animal phyla (see Chap. 2) has of course many variations. Normally and probably originally a large number of such membranes are formed either by the whole midgut epithelium, only in the anterior part, or only in the posterior part of the midgut. Therefore, the plural form *peritrophic membranes* should be used. Together such membranes surround the food mass in the midgut as a *peritrophic envelope*. Such an envelope may consist even of two different layers, each containing numerous peritrophic membranes of different origin and different texture of microfibrils; this has been observed for instance in the earwig, *Forficula auricularia* (Dermaptera) (Peters et al. 1979). The inner layer consists of membranes with a regular, orthogonal texture of microfibrils. Due to their formation several membranes stick together to form *aggregated peritrophic membranes* in which the orthogonal texture of microfibrils is in register. Only in specialized forms the ability to secrete peritrophic membranes is limited to a few rows of cells at the entrance of the midgut, as in most larvae of Diptera (Aubertot 1934; Zhuzhikov 1963; Peters 1976). In these cases, in fact, a single *peritrophic membrane* is secreted and therefore the singular form can be used. In all other cases the conventional use of the singular is wrong and should be replaced by the use of either the plural form or the term *peritrophic envelope*.

In some insects the end of the midgut is closed, for example in the ant lion, *Myrmeleon formicarius*, in the larvae of the honey-bee, *Apis mellifica*, and probably in almost all Hymenoptera Apocrita. In these cases a sac-like peritrophic envelope, consisting of several or numerous peritrophic membranes is formed which is called the *peritrophic sac*.

The lumen of the midgut is more or less divided into two compartments by the peritrophic membranes (Terra et al. 1979). This should be taken more into consideration by physiologists working on special features of digestion in insects, for instance, distribution of enzymes, water balance etc. The lumen of the peritrophic envelope, which is filled with the food mass, is called the *endoperitrophic space*. In this compartment the first phase of digestion, the degradation of larger food molecules by enzymes, occurs. Between the peritrophic membranes and the midgut epithelium there is a space which is called the *ectoperitrophic space*. This is a special compartment in which the second step in the degradation of food molecules by special enzymes takes place (Terra et al. 1979).

Morphological, and as far as we know, also biochemical asymmetries, occur at least in the peritrophic membranes of larvae of Diptera as well as in flies (Muscomorpha) (Smith 1968; Peters 1976, 1979; Becker 1977; Peters et al. 1983). In this case the lumen and the epithelium side must be distinguished. However, as these membranes tend to be thrown into deep folds, this is impossible in the lumen of the midgut. The orientation has to be determined in the formation zone of the cardia.

Despite the fact that Balbiani's concise and comprehensive definition exists, several modern authors regretted that there is no generally agreed-upon criterion for peritrophic membranes. Obviously, Balbiani's definition had been forgotten. For many years the chitin content was thought to be the only, generally accepted criterion of peritrophic membranes (Waterhouse 1953a). But in three phyla (Annelida, Mollusca and Chordata) peritrophic membranes have been observed which contain chitin and others which look alike but have no chitin content; those species which have chitin in their membranes belong to more ancestral groups than those in which the membranes are lacking chitin (Peters 1967a, 1968). The ability to synthesize chitin and to incorporate it into cuticles and other structures seems to be an ancestral feature of animals (Jeuniaux 1963). It seems to have been lost several times in the animal kingdom during evolution. Therefore, chitin cannot be used as the only criterion for peritrophic membranes.

Richards and Richards (1977) mentioned five possible characteristics for peritrophic membranes:

1. A positive chitin test, usually the chitosan test. However, it must be considered that the food may contain chitin in predaceous or cannibalistic species. Even mosquito larvae, which normally feed on detritus, may ingest large amounts of peritrophic membranes if they are reared in overcrowded cultures.
2. It has to be a membranous material which can be isolated from the food bolus or from fecal pellets.
3. A line separating the food from the epithelium in histological sections. In this case it has to be considered that sometimes precipitates may form during fixation. Very delicate structures have been observed by several authors in the midgut and Malpighian tubules of Hemiptera, Heteroptera as well as cicadas, but it has not yet been substantiated that these are really peritrophic membranes.
4. Any recognizable layer around the food produced by midgut cells. The difficulty is to distinguish between peritrophic membranes, mucous substances and the surface coat or glycocalyx.
5. Any membranous or filamentous secretion of midgut cells, regardless of whether or not it is concerned with food. In several species of Coleoptera, of the families Curculionidae and Ptinidae, the usual function of peritrophic membranes changes during development. These species use peritrophic membranes during the feeding phase to envelop food and food residues; after the cessation of feeding such membranes are used for cocoon formation (for a review, see Kenchington 1976; see also Sect. 3.3.3.1). *Murmidius ovalis*, a beetle of the family Cerylonidae, is a pest of stored products. In this species no peritrophic membranes could be detected as yet in the midgut of feeding larvae. But after the end of the feeding phase, the midgut epithelium starts to secrete chitin-containing threads which are obviously comparable to peritrophic membranes. These threads are extruded through the anus and are used for cocoon formation (Rudall and Kenchington 1973).

These remarks illustrate the variability of peritrophic membranes and the difficulties which arise for a clear-cut definition. Finally, Richards and Richards returned to the old definition given by Balbiani – without mentioning it – and added that most of these membranes contain chitin.

It is a peculiarity of biological systems that they have a core or a starting point which is clearly definable, and that they have a periphery in which the characteristic features are lacking (Hennig 1950, 1982). Only a sequence shows that species with ancestral features as well as species with highly specialized or reduced features belong to the same group. This is also true for the peritrophic membrane which can be regarded as an ancestral and widespread feature in the animal kingdom. As it can be assumed that the chitin content is also an ancestral feature (Jeuniaux 1963), and that chitin was present in peritrophic membranes originally, those forms which lack chitin in peritrophic coatings appear to be derived (several Mollusca, Annelida, Chordata, and the Hemiptera among Insecta; see Chap. 2).

Speculations on their origin during evolution led to the hypothesis that peritrophic membranes may be derived from the surface coat of midgut cells which is reinforced by chitin-containing microfibrils and delaminated frequently from the midgut cells in order to envelop food and food residues (Peters et al. 1983). In favor of this hypothesis is the fact that in larvae of a blowfly, *Calliphora erythrocephala*, only the lumen side of the single peritrophic membrane is provided with lectins which have a high specificity for mannose; moreover, it has mannose residues which probably belong to glycoproteins. The peritrophic membrane is formed only in the cardia by specialized cells, but not by cells of the midgut proper. The latter express neither the mannose-specific lectins nor mannose residues in their surface coat. Therefore, it seems possible that in this case the midgut cells could have lost their lectins during the course of evolution in favor of the peritrophic membrane (Peters et al. 1983).

These speculations could be corroborated by further investigations on the molecular composition and functions of the peritrophic membranes and the surface coat of the respective midgut cells.

Chapter 2

Occurrence of Peritrophic Membranes

Usually it is believed that peritrophic membranes are a special feature of insects and that they occur only in arthropods. However, this is the state of the art of the nineteenth century. In the meantime things have changed and this should be considered now by the authors of textbooks. The following specification is not an end in itself but intended as a stimulus to fill the gaps in our knowledge. Table 1 shows that it is easier to quote those phyla in which no peritrophic membranes have been found as yet. As chitin content is thought by many authors to be a valuable characteristic, this is mentioned in a separate column. Furthemore, + or – does not mean that all species of the respective group have chitin or no chitin in their peritrophic membranes. More detailed information will be given in the following.

1. Plathelminthes
No peritrophic membranes have been found in this group. Nothing is mentioned in the literature. The following species were investigated by Peters (1968): *Stenostomum leucops, Dugesia lugubris, Fasciola hepatica, Dicrocoelium dendriticum* and *Haematoloechus variegatus*.

2. Nemertini
Cerebratulus sp. and *Lineus ruber* seem to have no peritrophic membranes. In the latter species the feces are surrounded by a considerable amount of slime which seems to be supplied by the epidermis and not by the midgut epithelium; the slime is rather resistant to 40% KOH, but does not contain chitin.

3. Nemathelminthes
In the nematode species *Ascaris lumbricoides* and *Leidynema appendiculata* no peritrophic membranes were observed. Nothing is known about the occurrence of peritrophic membranes in other groups of the Nemathelminthes.

4. Kamptozoa
Pedicellina cernua had no peritrophic membranes. There are no reports in the literature.

5. Priapulida
The midgut of *Halicryptus spinulosus* contains relatively large pellets of food residues which are surrounded by several or numerous membranes. These give a strong positive chitosan reaction. The peritrophic membranes contain microfibrils which are arranged in a random texture (Fig.1).

Table 1. Occurrence of peritrophic membranes

Phylum or class	Petritrophic membranes	Chitin in PM
Plathelminthes	−	
Nemertini	−	
Nemathelminthes	−	
Kamptozoa	−	
Priapulida	+	+
Sipunculida	+	−
Echiurida	−	
Mollusca	−	
Solenogastres	?	
Caudofoveata	?	
Polyplacophora	+	+
Gastropoda	+	+
Scaphopoda	−	
Lamellibranchia = Bivalvia	−	
Cephalopoda	−	
Annelida		
Polychaeta	+	+
Archiannelida	+	+
Myzostomida	+	?
Oligochaeta	+	+
Hirudinea	−	
Onychophora	+	+
Linguatulida	?	
Tardigrada	?	
Arthropoda		
Merostomata	+	+
Arachnida	+	+
Pantopoda	−	
Crustacea	+	+
Chilopoda	+	+
Diplopoda	+	+
Pauropoda	?	
Symphyla	?	
Insecta	+	+
Tentaculata		
Phoronidea	?	
Bryozoa	+/−	
Brachiopoda	+	+
Hemichordata	+	−
Echinodermata		
Crinoidea	?	
Holothuroidea	+	−
Echinoidea	+	−
Asteroidea	?	
Ophiuroidea	?	
Chaetognatha	?	
Chordata		
Tunicata	+	+
Acrania	+	?
Agnatha	+	?
Verterbrata	+	−

6

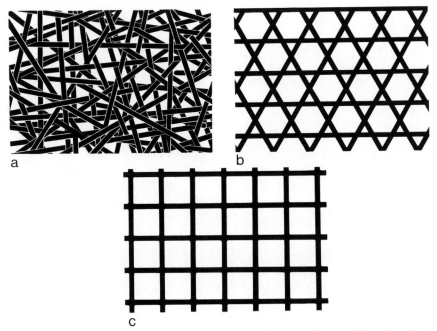

a   b

c

Fig. 1. Schematic representation of the three types of textures of chitin-containing microfibrils in peritrophic membranes. *Upper left*, random texture; *upper right*, honeycomb or hexagonal texture which is usually combined with layers of random texture; *below*, grid or orthogonal texture (After Peters 1968)

6. Sipunculida

Phascolosoma vulgare and *Ph. gouldi* had no membranous material around their gut contents, whereas *Sipunculus nudus* secreted a thin, but resilient membrane which was dissolved in 40% KOH. The cuticle of *Sipunculida* contains also no chitin (Jeuniaux 1963).

7. Echiurida

The midgut of *Bonellia viridis* and *Echiurus echiurus* is filled with small fecal pellets. However, these are not held together by a recognizable membrane, but by the stickiness of the food residues.

8. Mollusca

Polyplacophora: All species which were investigated had peritrophic membranes: *Lepidochiton cinereus*, *Acanthochites fascicularis*, *Katharina tunicata* (Fig. 2) and *Cryptochiton stelleri* had several delicate membranes in their midgut which surrounded food residues. The membranes around the fecal pellets of *Lepidochiton cinereus* were too fragile for the chitosan test, but the membranes of the other species were chitosan positive. However, prolonged hydrolysis for 5–7 days with 40% KOH

7

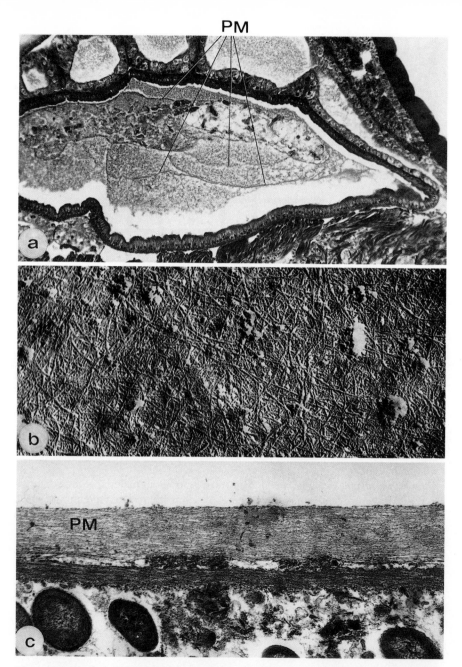

Fig. 2. *a* In the midgut of *Acanthochites fascicularis* (Mollusca, Polyplacophora) peritrophic
membranes (*PM*) envelop food and food residues. Light micrograph of a longitudinal section
stained with molybdate hematoxylin after Dobell; 200x. *b* Isolated peritrophic membranes
of *Ischnochiton* sp. (Mollusca, Polyplacophora) after shadow-casting with platinum. They
contain microfibrils which are arranged at random; 37 000x. *c* The peritrophic membranes
(*PM*) of the pond snail, *Limnaea stagnalis* (Gastropoda, Basommatophora) consist of slime
which contains no chitin and no microfibrils; 27 500x. *a,b* After Peters (1968)

at 60 °C was necessary. Electron microscopy showed that these peritrophic membranes contain microfibrils arranged in a felt-like texture.

Gastropoda: *Patella coerulea* is able to secrete delicate chitin-containing peritrophic membranes which proved to contain microfibrils. The other species which were investigated secreted a mucous and viscous material around the fecal string in the gut. But in the electron microscope this material looked more like a membrane than a shapeless mucous secretion in the pond snail, *Limnaea stagnalis* (Fig. 2c).

Such mucous material which contained no chitin was observed in *Crepidula fornicata, Pleurobranchea meckeli, Limnaea stagnalis, Cepaea hortensis, Arianta arbustorum, Biomphalaria glabrata* and *Limax flavus*.

Scaphopoda, Lamellibranchia and Cephalopoda contained no membranous material in their midgut. The following species were investigated: *Dentalium dentale, Nucula nucleus, Anodonta cygnea, Spaerium corneum, Sepia officinalis, Loligo pealei* and *Octopus vulgaris*.

9. Annelida

Polychaeta: In this group more species have been investigated than in the preceding phyla. Therefore, an easy to survey synopsis is given in Table 2. A few remarks should be added.

Arenicola marina: As this species feeds on sand enriched with small animals and detritus, no peritrophic membranes are expected. However, Vierhaus (1971) found in the midgut and around freshly formed feces membranous material which gave a weak, but positive chitosan test.

Amphitrite ornata: Dales described in 1955 a "peritrophic membrane" in the anterior part of the gut which was chitosan positive. However, it proved to be a cuticular lining with a random arrangement of microfibrils embedded in a matrix (Dales 1967). In the midgut of preserved specimens, which was filled with abundant food, Vierhaus (1971) could detect no peritrophic membranes; but in the hindgut, the contents were surrounded by a mucous envelope.

Neoamphitrite figulus: Dales and Pell (1970) reported that a peritrophic membrane lines the muscular gizzard or hind stomach, and to a lesser extent the secretory fore stomach region of the gut. It was reported to contain the β-form of chitin like the chaetae. A reinvestigation by Rudall and Kenchington (1973) revealed that it did not contain β- but α-chitin. Its thickness varies considerably and it may be absent in worms which have fasted for 1–2 days. It can be removed from the epithelium as a gelatinous tube. Obviously, this is not a peritrophic membrane as these are formed in other Annelida by the midgut epithelium in the form of thin lamellae. The secretion product of the gizzard of *Neoamphitrite* resembles the cuticle of earthworms which also contains chitin and which is constantly regenerated as it is abraded by food material. It serves as a matrix for enzymes like amylases and proteases (Peters and Walldorf 1986a). This cuticle closely resembles the gastric shield which has been found in species from all groups of Mollusca (Owen 1966).

Table 2. Peritrophic membranes in Annelida[a]

	PM	MF	Chitin	Author
Polychaeta				
Aphrodite aculeata	+	+	+	
Harmothoe sarsi	+		−	Vierhaus (1971)
Nereis diversicolor	+	+		
Nereis virens	+	+	+	
Platynereis dumerilii	+		−	Vierhaus (1971)
Nephthys hombergi	+		+	
Ophryotrocha sp.	+		+	
Arenicola marina	−			
	+		+	Vierhaus (1971)
Maldane sp.	+		−	Vierhaus (1971)
Pectinaria koreni	−			Vierhaus (1971)
Melinna palmata	(+)		−	Newell and Baxter (1937)
Melinna cristata	?			Wirén (1885)
Terebella lapidaria	−			Sutton (1957)
Amphitrite johnstoni	+			Dales (1955)
Amphitrite ornata	(+)			Vierhaus (1971)
Nicolea venustula	+		+	Vierhaus (1971)
Lanice conchilega	+	+	+	
Sabella sp.	+	+	+	Vierhaus (1971)
Spirographis spallanzani	+	+	+	
Archiannelida				
Dinophilus gyrociliatus				
Myzostomida				
Myzostomum sp.	+			
Oligochaeta				
Aeolosoma hemprichi	−			Vierhaus (1971)
Dero obtusa	+		+	
Nais sp.	(+)		−	Vierhaus (1971)
Tubifex tubifex	(+)	−	−	Vierhaus (1971)
Limnodrilus hoffmeisteri	(+)	−	−	Vierhaus (1971)
Lumbriculus variegatus	+			Vierhaus (1971)
Enchytraceus albidus	+		−	
Eisenia foetida	+			van Ganseu-Semal (1962)
	+		+	Vierhaus (1971)
Dendrobaena octaedra	+			Vierhaus (1971)
Dendrobaena subrubicunda	+	+	+	Vierhaus (1971)
Dendrobaena rubida	+		+	Vierhaus (1971)
Octolasium cyaneum	+	+		Vierhaus (1971)
Octolasium lacteum	+		+	Vierhaus (1971)
Allolobophora caliginosa	+	+	+	Vierhaus (1971)
Allolobophora chlorotica	+	+	+	Vierhaus (1971)
Allolobophora longa	+			Vierhaus (1971)
Allolobophora jenensis	+	+		Vierhaus (1971)
Allolobophora rosea	+			Vierhaus (1971)
Lumbricus terrestris	+	+	+	Vierhaus (1971)
Lumbricus rubellus	+	+	+	Vierhaus (1971)
Hirudinea				
Glossiphonia complanata	−			Vierhaus (1971)
Theromyzon tessulatum	−			Vierhaus (1971)
Hemiclepsis marginata	−			Vierhaus (1971)
Hirudo medicinalis	−			Vierhaus (1971)
Haemopis sanguisuga	+	+	+	Vierhaus (1971)
Erpobdella octoculata	−			
Erpobdella octoculata	+		+	Vierhaus (1971)
Erpobdella testacea	+	+	+	Vierhaus (1971)

[a]PM, peritrophic membranes; MF, microfibrils were found with the electron microscope; chitin results of the chitosan test.

10

Oligochaeta: Thirteen species of earthworms have been investigated as yet. In most of these chitin could be detected with the chitosan test. The staining intensity differed from one species to another. In several species microfibrils forming a random texture could be observed with the electron microscope (Fig. 27). Estivating specimens of *Allolobophora chlorotica*, which had no food in their gut, contained no peritrophic membranes (Vierhaus 1971).

Hirudinea: Blood-sucking species had no peritrophic membranes, whereas the predators, e.g. *Haemopis sanguisuga* and *Erpobdella octoculata*, secreted membranes which surrounded the food residues in the midgut; these contained microfibrils and yielded a positive chitosan test (Vierhaus 1971).

10. Onychophora

Manton and Heatley (1937) were the first to observe peritrophic membranes in a species of Onychophora, *Peripatopsis moseleyi*. They described the formation, rhythm of secretion, and enclosure of uric acid crystals and its daily removal (see Chap. 6). Waterhouse found such membranes in *Ooperipatus paradoxus*. De Mets et al. (1965) tried to elucidate the fine structure of these membranes, but they could not distinguish between peritrophic membranes and food residues.

11. Linguatulida

No peritrophic membranes have been found as yet.

12. Tardigrada

As most tardigrades are observed soon after their emergence from a cyst and before they have filled their midgut with food, no peritrophic membranes have been described as yet in this group.

13. Arthropoda

Merostomata: In *Limulus polyphemus* a large number of peritrophic membranes form a peritrophic envelope which surrounds the feces in the midgut and hindgut. De Mets (1962) found that the microfibrils form a hexagonal or honeycomb texture. Peters (1967a) established the presence of chitin in these peritrophic membranes.

Arachnida: Scorpions, like *Euscorpius italicus*, *Buthus occitanus* and *Centruroides* sp., are able to secrete peritrophic membranes. Tulk (1843) was the first to describe peritrophic membranes which surround the large fecal pellets of harvestmen, *Phalangium opilio* (Opilionida). His findings were confirmed by Milne-Edwards (1857–1865), Plateau (1876) and Kästner (1933), and by the light and electron microscopic investigations of Peters (1967b) and Becker and Peters (1985) which showed the presence of chitin and an irregular as well as a hexagonal texture of microfibrils in the membranes (Figs. 29, 30). Grainge and Pearson (1965) observed peritrophic membranes in *Leiobunum rotundum*, and Peters (1967b) found in this species and in *Nelima aurantiaca* that chitin is present in these membranes. Although spiders have extraintestinal digestion and feed on fluid food, all species which were investigated as yet had peritrophic membranes in their midguts: *Agelena labyrinthica*, *Argyroneta aquatica*, *Araneus umbraticus* (Plateau 1876), *Araneus diadema-*

tus, Linyphia triangularis (van der Borght 1966), and *Tegenaria derhami* and *Pholcus phalangoides* (Peters 1968). In *Araneus diadematus* the author could show the presence of chitin in such membranes. This is another example that fluid feeders can have peritrophic membranes, as has been shown already by several authors, especially Waterhouse (1953a) for many species of insects (Lepidoptera and Diptera) (see also Peters 1969).

"Acari": Samson (1909) described for the first time a peritrophic membrane in a tick, *Ixodes ricinus*. Six days after a blood meal, an "intima" appeared on the apical cell surface of the midgut cells of females. It increased in thickness during egg deposition and finally detached when the blood had been resorbed. The appearance of the detached intima reminded Samson of the peritrophic membrane of insects. Moreover, she assumed that lifting of the intima might cause the disintegration of the midgut epithelium at the end of blood digestion. Aeschlimann (pers. comm.) confirmed the presence of a peritrophic membrane covering the midgut cells of nymphs and females of *Ixodes ricinus* already 26 and 22 h after attachment. It is a single, homogeneous, uneven layer with a spongy appearance, low electron density and a thickness of about 60–120 nm. After detachment from the midgut cells, which occurs 5–6 days after the beginning of a blood meal, its thickness increased five times that before lifting. Both investigations show that peritrophic membranes are formed in ticks as well as in adult female mosquitoes only after a blood meal. Gern et al. (1990) reported that the culture medium BSK-II for spirochaetes is able to induce the formation of a peritrophic membrane in *Ixodes ricinus*. Rudzinska et al. (1982) could demonstrate electron microscopically peritrophic membranes in the gut of feeding larvae, nymphs and adults of another species, *Ixodes damini*. Peritrophic membranes were observed also in a species of the Argasidae, *Ornithodorus moubata*, by Grandjean (1984).

Nothing can be found in the literature about such membranes in the midgut of species of the other large group of Parasitiformes, the Gamasidae.

However, in Acaroidea, several findings of peritrophic membranes have been recorded. Hughes (1950) found them in *Tyroglyphus farinae*. Zachvatkin (1959) mentioned already that food balls in the gut of Acaroidea were held together by "a very thin film (to a certain extent analogous with the peritrophic membranes formed by insects)". Prasse (1967) found such membranes in *Caloglyphus berlesei*, Wharton and Brody (1972) in *Dermatophagoides farinae*, and Vijayambika and John (1975) in fish mites, *Lardoglyphus konoi*; in the latter case, the colon is said to be responsible for the formation of these membranes. Wharton and Brody (1972) observed that the incoming food was wrapped in a sac-like membrane which was formed at the entrance of the midgut. This peritrophic membrane was 1–2 μm thick. Finally, this membrane delaminated from the formation zone. The so-called food ball was transported to the hindgut where its size decreased eight fold, probably in connection with water resorption. Several food balls were compacted to form a fecal pellet which seemed to be held together by mucus.

The most intensive investigation was done by Akimov (1980) who observed peritrophic membranes in the midgut of 15 genera of Acaridae and Glyciphagidae: *Acarus siro, Tyrophagus putrescentiae, Aleuroglyphus ovatus, Kuzinia laevis, Rhizoglyphus echinopus, Caloglyphus berlesei, Schwiebea rossica, Histiogaster bacchus,*

Thyreophagus entomophagus, *Chortoglyphus arcuatus*, *Glycyphagus domesticus*, *Coleochaeta molitor*, *Gohieria fusca*, *Carpoglyphus lactis* and *Lardoglyphus konoi*. Oribatids are also able to secrete peritrophic membranes which surround the fecal pellets. This could be shown by Dinsdale (1975) in *Phthiracarus* sp. Therefore, undoubtedly peritrophic membranes also occur often in the "Acari".

Pantopoda: Nothing is known about the occurrence of peritrophic membranes in this group. Random tests with *Pycnogonum pusillum* and *Nymphon* sp. gave no results as the guts of the specimens were empty.

Crustacea: A large number of species from different groups have been investigated. The results show that peritrophic membranes appear to be an ancestral, almost constant feature of the Crustacea which might be abandoned only in some specialized species. However, it might be that some of the latter species which were investigated were only lacking peritrophic membranes at the time of investigation. Table 3 gives a survey of the occurrence of peritrophic membranes in Crustacea.

Darwin was the first to describe peritrophic membranes in Crustacea. He found transparent membranes which enveloped the feces, and he assumed therefore that these membranes consisted of chitin. However, Waterhouse (1953b) was the first to demonstrate definitely that the peritrophic membranes of a phyllopod, *Simona* sp., contains chitin. Georgi (1969b) could show in his extensive comparative studies that in Crustacea the chitin-containing microfibrils are arranged either in a random or a hexagonal or honeycomb texture. An orthogonal texture was found only in *Chirocephalus grubei* and in *Triops cancriformis* by Georgi (1969a) and in *Leptestheria dahalacensis* by Schlecht (1977, 1979). Gauld (1957) reported that peritrophic membranes are formed in the posterior part of the midgut of a copepod, *Calanus* sp. These are easily visible by staining with Congo red. Chitin could not be demonstrated in these membranes, perhaps because they are too small and delicate. Gauld remarked that the peritrophic membranes surround the fecal pellets and thus ensure that the indigestable remains of the food are rapidly removed from the water in which the animals are feeding. Gauld assumed that the fecal pellets sink and that thus the food residues are not ingested over and over again by the same individual. In the past 30 years a vast amount of literature has accumulated which shows that these fecal pellets are of considerable ecological importance (see Sect. 6.7).

In Crustacea peritrophic membranes may be formed either by the whole midgut epithelium (Schlecht 1977, 1979) or by the anterior part of the midgut, as reported for *Marinogammarus* sp. by Martin (1964), or by the posterior part of the midgut, as observed in *Calanus* sp. by Gauld (1957).

Chilopoda: Unexpectedly, *Lithobius forficatus* was among the first animals in which peritrophic membranes were observed (Plateau 1876; Gibson-Carmichael 1885). Plateau found them also in *Himantarium* sp. Balbiani (1890) found these membranes in the midgut of *Cryptops punctatus* and *C. hortensis* as well as in *Lithobius forficatus*. Waterhouse (1953) demonstrated the presence of chitin in the peritrophic membranes of *Cormocephalus aurantipes* and *Allothereua maculata*. Peters (1968) could show that the peritrophic membranes of *Lithobius forficatus* and *Scolopendra cingulata* contain microfibrils and chitin (Fig. 3a). Rajulu (1971) described the fine structure and chemical composition of the peritrophic membranes

Table 3. Occurrence of peritrophic membranes in Crustacea[a]

Group and species	PM	MF	Chitin	Author
Anostraca				
Artemia salina	+			Reeve (1963)
Artemia salina	+		+	Peters (1968)
Artemia salina	+			Snyder and Wolfe (1980)
Chirocephalus grubei	+	+		Georgi (1969a)
Phyllopoda				
Triops cancriformis	+	+		Georgi (1969a)
Lepidurus apus	–			Georgi (1969a)
Leptestheria dahalacensis	+	+	+	Schlecht (1977, 1979)
Daphnia magna	+			Chatton (1920)
Daphnia magna	+	+	+	Schlecht (1977, 1979)
Daphnia pulex	+			Aubertot (1932)
Daphnia pulex	+	+		Georgi (1969a)
Daphnia pulex	+			Güldner (1969)
Daphnia pulex	+	+		Schultz and Kennedy (1976)
Daphnia pulex	+	+	+	Schlecht (1977, 1979)
Diaphanosoma brachyurum	+	+		Schlecht (1977)
Simona sp.	+		+	Waterhouse (1953a)
Ostracoda				
Iliocypris gibba	+	+		Waterhouse (1953a)
Cypicerus affinis	+	+		Waterhouse (1953a)
Copepoda				
Cyclops sp.	+			Farkas (1922)
Calanus sp.	+			Gauld (1957)
Cyclops albidus	+			Georgi (1969a)
Caligus curtus	?			Georgi (1969a)
Branchiura				
Argulus foliaceus	?			Georgi (1969a)
Cirripedia				
Balanus balanoides	+			Darwin (1854)
Balanus balanoides	+			Rainbow and Walker (1977)
Balanus hameri	+			Rainbow and Walker (1977)
Balanus improvisus	+			Rainbow and Walker (1977)
Balanus porcatus	+	+	+	Georgi (1969a)
Balanus nubilus	+	+	+	Georgi (1969a)
Chthamalus stellatus	+			Darwin (1854)
Lepadidae	+			Darwin (1854)
Lepas anatifera	+	+	+	Georgi (1969a)
Chelonibia patula	+	+	+	Georgi (1969a)
Mitella spinosus	+			Batham (1945)
Mitella polymerus	+	+	+	Georgi (1969a)
Malacostraca				
Stomatopoda				
Squilla mantis	+	+		Georgi (1969a)
Euphausiacea				
Euphausia superba	?			Georgi (1969a)
Decapoda				
Penaeus setiferus	+	+	+	Georgi (1969a)

Table 3. (Continued)

Group and species	PM	MF	Chitin	Author
Metapenaeus bennettae	+			Dall (1967)
Pandalus montagui	+			Forster (1953)
Pandalina brevirostris	+			Forster (1953)
Hippolyte varians	+		+	Forster (1953)
Spirontocharis cranchii	+			Forster (1953)
Spirontocharis pusiola	+			Forster (1953)
Athanas nitescens	+			Forster (1953)
Processa canalicula	+			Forster (1953)
Leander serratus	+		+	Forster (1953)
Leander squilla	+			Forster (1953)
Palaemonetes varians	+	+		Georgi (1969)
Crangon vulgaris	+			Forster (1965)
Philocheras trispinosus	+			Forster (1953)
Palinurus vulgaris	+	+	+	Georgi (1969)
Homarus gammarus	+	+	+	Georgi (1969)
Nephrops norvegicus	+	+	+	Georgi (1969)
Orconectes limosus	+	+	+	Peters (1968)
Orconectes limosus	+	+	+	Georgi (1968)
Astacus leptodactylus	+	+	+	Georgi (1969)
Eupagurus bernhardus	+	+	+	Georgi (1969)
Coenobita jousseaumei	+	+	+	Georgi (1969)
Galathea intermedia	+	+	+	Georgi (1969)
Calappa granulata	+	+	+	Georgi (1969)
Hyas araneus	+	+	+	Georgi (1969)
Carcinus maenas	+	+	+	Peters (1968)
Carcinus maenas	+	+	+	Georgi (1968)
Cancer pagurus	+	+	+	Georgi (1969)
Cancer magister	+			Holliday et al. (1980)
Xantho hydrophilus	+	+	+	Holliday et al. (1980)
Eriphia spinifrons	+	+	+	Holliday et al. (1980)
Eriocheir sinensis	+	+		de Mets (1962)
Eriocheir sinensis	+	+	+	Georgi (1969)
Mysidacea				
Praunus flexuosus	+	+		Georgi (1969)
Isopoda				
Asellus aquaticus	?			Georgi (1969)
Porcellio scaber	?			Georgi (1969)
Armadillidium vulgare	–			Vernon et al. (1974)
Mesidotea entomon	+	+	+	Vernon et al. (1974)
Aega psora	+	+	+	Vernon et al. (1974)
Rocinela dumerili	+	+	+	Vernon et al. (1974)
Limnoria tripunctata	+			Zachary et al. (1983)
Amphipoda				
Marinogammarus sp.	+			Martin (1964)
Rivulogammarus pulex	+	+	+	Peters (1968)
Rivulogammarus pulex	+	+		Georgi (1969)
Gammarus lacustris	+			Lautenschlager et al. (1978)
Talitrus saltator	+	+		Lautenschlager et al. (1978)
Elasmopus rapax	+	+	+	Lautenschlager et al. (1978)

[a]PM, peritrophic membranes; MF, microfibrils were found with the electron microscope; chitin results of the chitosan test.

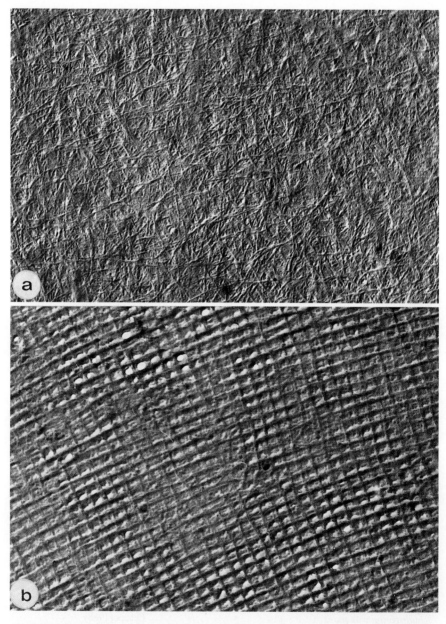

Fig. 3a,b. Isolated peritrophic membranes of Myriapoda after shadow-casting with platinum; 37 000x. *a Lithobius forficatus* (Chilopoda). *b Spirobolus* sp. (Diplopoda) (After Peters 1968)

of *Ethmostigmus spinosus*. They contain chitin and the microfibrils form an orthogonal texture. In Pauropoda, *Allopauropus vulgaris* and *Pauropus huxleyi* (Zanger unpubl. results) and Symphyla no peritrophic membranes have been demonstrated as yet.

Diplopoda: Plateau (1876) had the opinion that herbivores like *Julus* species are unable to secrete peritrophic membranes. However, Schneider (1887) could show that peritrophic membranes exist in *Julus* sp. Hefner (1929) found them in *Parajulus impressus*, Miley (1930) in *Euryurus erythropygus*, Bowen (1968) in *Floridobolus penneri* and *Narceus gordanus*, and Nunez and Crawford (1977) reported them from the midgut of the desert millipede *Orthoporus ornatus*. Mason and Gilbert (1954) investigated millipedes from different systematic groups and always found peritrophic membranes with chitin in them: *Polyxenus lagurus*, *Tachypodoiulus niger*, *Ophyiulus pilosus*, *Julus scandinavius*, *Cylindroiulus punctatus*, *Glomeris marginata*, *Polydesmus angustus*, *Polymicrodon polydesmoides* and *Choneiulus palmatus*. Only in the small *Polyxenus lagurus* it was impossible to apply the chitosan test. Peters (1968) found peritrophic membranes with chitin microfibrils arranged in an orthogonal texture in *Glomeris marginata*, *Julus* sp. and *Spirobolus* sp. (Fig. 3b).

Insecta: Insects were the first, and with respect to the occurrence of peritrophic membranes, most intensely investigated group of animals. The reviews by Aubertot (1934), Peters (1969) and Richards and Richards (1977) give a rather comprehensive picture of the subject. Therefore, only a few comments will be given here.

There are a few orders in which peritrophic membranes seem to be absent: Phthiraptera, Psocoptera, Thysanoptera, Diploglossata, Zoraptera, Strepsiptera, Rhaphidioptera and Megaloptera. Adult fleas (Siphonaptera) have no peritrophic membranes, whereas their larvae are able to secrete a continuous, tube-like membrane at the beginning of the midgut. According to earlier reports, and subsequently to textbooks, peritrophic membranes are said to be lacking in several orders and smaller taxa. However, more recently such membranes were detected.

Diplura: In a larger species of the Diplura, *Heterojapyx evansi*, lamellar peritrophic membranes are secreted by the whole midgut epithelium (Waterhouse 1953a).

Protura: Dallai (1974) found such membranes in *Eosentomon* sp. and *Acerentomon* sp.

Zygentoma: Earlier reports of peritrophic membranes in the common silverfish, *Lepisma saccharina* (Aubertot 1932a,b) were confirmed by Waterhouse (1953a), who found chitin in them, by Peters who detected microfibrils in them, and by Larsson (1973) who investigated light microscopically the histology of the alimentary tract. Waterhouse (1953a) found peritrophic membranes in *Ctenolepisma longicaudata*.

Archaeognatha: Peters (1969) described peritrophic membranes with microfibrils arranged in a random texture from *Lepismachilis* sp.

Ephemeroptera: Adults do not consume food, whereas larvae do so and have typical peritrophic membranes (Figs. 4a and 22).

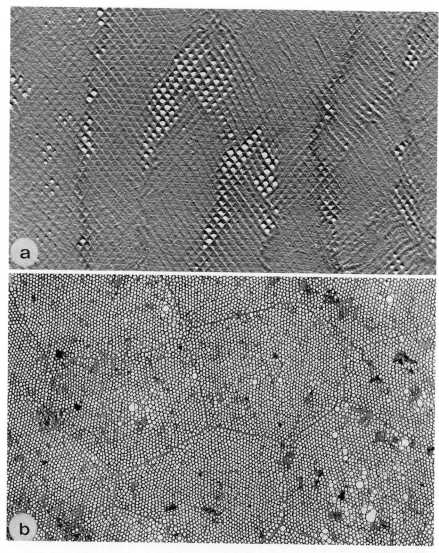

Fig. 4a,b. Isolated peritrophic membranes after shadow-casting with platinum. *a* Peritrophic membrane of a mayfly larva, *Cloeon dipterum* (Insecta, Ephemeroptera). In this case the matrix seems to be preserved and partially fills the orthogonal network of chitin-containing microfibrils; 18 500x. *b* Peritrophic membrane of the cricket, *Acheta domestica* (Insecta, Saltatoria), with indications of midgut cell borders; 3800x (After Peters 1968)

Saltatoria: Waterhouse (1953) confirmed earlier findings by Cuénot (1895) in *Gryllotalpa gryllotalpa* and in *G. coarctata*. Four long plate-like extensions of the foregut extend through the short midgut into the hindgut. Although the contents of the midgut are fluid in nature, the midgut cells have not entirely lost their ability to secrete peritrophic membranes. Small fragments of membrane occurred near the surface of the cells and gave a positive test for chitin. Aubertot (1932a,b) observed already these membranous fragments which had originated from the midgut epithelium, but as he had no reliable test for chitin, he did not recognize their significance. The texture of microfibrils in the peritrophic membranes of the cricket is shown in Fig. 4b.

Embiodea: Waterhouse (1953a) reported from an unidentified species the presence of chitin-containing peritrophic membranes which were secreted at the entrance of the midgut. Peters (1969) found also such membranes in the midgut of *Embia* sp.

Hemiptera (Heteroptera and Homoptera): Despite the belief that peritrophic membranes protect the midgut epithelium from abrasive food material and that blood- or sap-feeding animals do not need them, careful observation revealed that even in Hemiptera and Homoptera a peculiar type of membrane occurs in the midgut. Light microscopic observations showed that at least in a few species of Hemiptera peritrophic membranes might be present. Sutton (1951) reported that Corixidae have delaminated chitin-containing peritrophic membranes, and Waterhouse (1953a) found such membranes in *Nezara viridula* (Pentatomidae); but in the latter species it was uncertain whether chitin is present. Parsons (1957) reported the presence of peritrophic membranes in *Ranatra fusca*. Electron microscopic investigations showed the presence of peculiar membrane systems in the midgut of Heteroptera, e.g. *Triatoma infestans* (Burgos and Gutiérrez 1976; Gutiérrez and Burgos 1978), *Rhodnius prolixus* (Lane and Harrison 1979; Bauer 1981; Billingsley and Downe 1983), *Nepa cinerea* (Andries and Torpier 1982), *Oncopeltus fasciatus* (Baerwald and Delcarpio 1983) (Fig. 5), *Lygaeus pandurus* (Tieszen et al. 1985, 1986), *Cimex lectularius, Rhodnius prolixus, Triatoma infestans, T. maculata, Anthocoris nemorum* and *Leptopterna dolobrata* (Chaika 1979), and in Homoptera in the cicadas *Cicadella viridis* (Gouranton and Maillet 1965), *Fulgora candelaria* (Marshall and Cheung 1970) and *Phylloscelis atra* (Reger 1971).

In another species of cicadas, *Euscelidius variegatus* (Jassidae), Munk (1967a,b) described the delamination of delicate peritrophic membranes from the epithelium of the filter chamber or posterior midgut and the cryptonephridium or Malpighian tubules, but the micrographs do not show a parallel to the double membranes of Hemiptera.

In aphids no such membranes have been described as yet.

Marshall and Cheung called these membranes a "plexiform surface coat", whereas others called it "extracellular membrane layers (ECML)". Most authors assume that this is a modified form of peritrophic membranes. Cytochemical investigations proved that these structures contain at least glycophospholipids, proteins, carbohydrates and hydrolytic enzymes. Therefore, they may act as a carrier for immobilized enzymes. Their formation is described in Section 4.3.2.

Coleoptera: The texture of microfibrils in the peritrophic membranes of the larvae of *Oryctes nasicornis* and *Leptinotarsa decemlineata* is shown in Figs. 5–7. Peritrophic membranes are said to be absent in groups with extraintestinal digestion, e.g. in Carabidae and Dytiscidae (Rungius 1911; Bess 1935; Wigglesworth 1930, 1972), but Aubertot was able to find membranous material in *Calosoma inquisitor*, three *Carabus* species, *Cicindela campestris, Dytiscus* sp. and *Cybister* sp. Water-

19

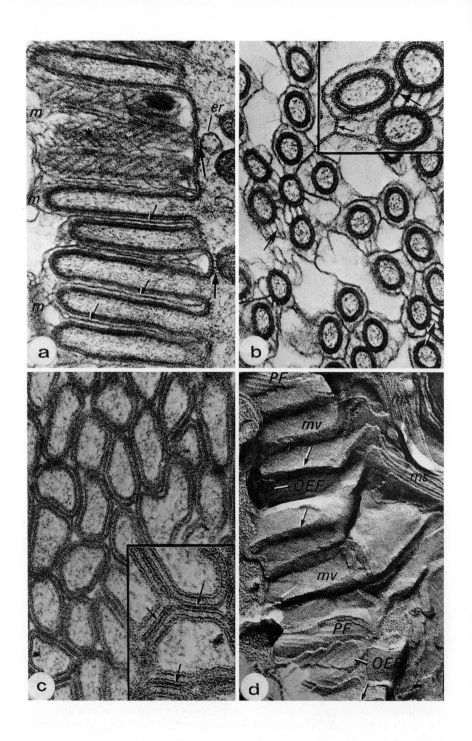

house (1953a) confirmed this by the observation of chitin-containing peritrophic membranes in the posterior half of the midgut of *Calosoma schayeri*. More recently, Cheeseman and Pritchard (1984) investigated two species of Carabidae with distinct feeding habits. Adult *Scaphinotus marginatus* liquefy their prey preorally and ingest this fluid, whereas those of *Pterostichus adstrictus* ingest particles of food, including indigestible components. In both species peritrophic membranes were observed which contain chitin as revealed by the chitosan reaction. Although specimens were fixed at various times after feeding, it could not be substantiated whether in these species the formation of peritrophic membranes is linked to the uptake of food into the midgut.

It is always quoted that, according to Möbuβ (1897), larvae of *Anthrenus* species do not form peritrophic membranes. But Waterhouse (1952, 1953a) could show that chitin-containing peritrophic membranes occur in the posterior half of the midgut in larvae of *Anthrenus vorax* and other Dermestidae. Sinha (1958) denied the presence of peritrophic membranes in adult *Tribolium castaneum*, however, Ameen and Rahman (1973) found them in both larval and adult stages of this species.

Hymenoptera: Loele (1914) and von Dehn (1933) reported that peritrophic membranes are absent in adult ants. But they investigated only the specialized genera *Formica, Myrmica* and *Camponotus*. However, Waterhouse (1953a) found well-formed peritrophic membranes in the posterior two-thirds of the midgut of two species of the more primitive Ponerinae, *Myrmecia nigriceps* and *Promyrmecia pilosula*; no membranes could be detected in *Iridomyrmex detectus* and *Camponotus consubrinus*. Wasps, bumblebees and honeybees possess numerous peritrophic membranes with textures of chitin microfibrils (Figs. 24, 36).

Mecoptera: Grell (1938) found no peritrophic membranes during his careful investigations of the midgut of adults and larvae of *Panorpa communis*, but Waterhouse (1953a) reported chitin-containing peritrophic membranes in adult *Harpobittacus australis*.

Lepidoptera: Schneider's (1887) opinion that the function of peritrophic membranes is to protect the midgut from hard and sharp particles in the food is nearly

←――

Fig. 5a-d. In bugs, in this case in *Rhodnius prolixus*, membranes delaminate from the microvilli of the midgut epithelium and accumulate in the lumen. These membranes were called "plexiform membranes" or "extracellular membrane layers". *a* Longitudinal section through microvilli with delaminating membranes (*arrows*). In obliquely sectioned parts (*asterisk*) the membrane (*m*) has a network appearance. Between the bases of microvilli the double membrane is connected with mitochondria and endoplasmic reticulum (*er*) by fibrillae (*thick arrows*); 48 400x. *b* Cross-sections of microvilli show outer membranes (*thick arrows* protruding into the lumen of the midgut or connecting each other); 48 500x. The *inset* reveals that both the outer membrane and its projections have the same characteristic appearance; 85 700x. *c* Cross-sections of microvilli from the anterior midgut demonstrate the close association of the outer membranes which form gap junction-like relationships (*arrows* in *inset*); 74 500x; *inset* 138 000x. *d* Freeze fracture replica of microvilli (*mv*); stacks of membrane (*ms*) on the lumen side. *Arrows* point to rows of particles within the membranes. Rows of particles can also be seen on both sides of the microvillus fractures (*PF, OEF*); 52 500x (After Lane and Harrison 1979)

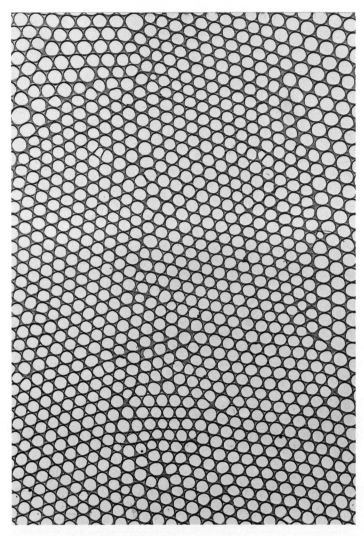

Fig. 6. Isolated peritrophic membrane of a larvae of *Oryctes nasicornis* (Coleoptera) after shadow-casting with platinum. This specimen displays a hexagonal texture with many irregularities 18 000x (After Peters 1969)

always cited in textbooks which deal with this problem. Consequently, it has been assumed that peritrophic membranes are unnecessary and therefore absent in fluid-feeding insects. As has been shown in the preceding part, this may be true for several insects, but is not a general rule, as could be shown already by Waterhouse (1953a). The majority of adult Lepidoptera feed only on liquids, if at all. Before Waterhouse started his comprehensive investigations, there were only two references demonstrating that peritrophic membranes exist in Lepidoptera: Petersen (1912) found these membranes in *Pieris rapae*, and Aubertot (1932a,b) described them

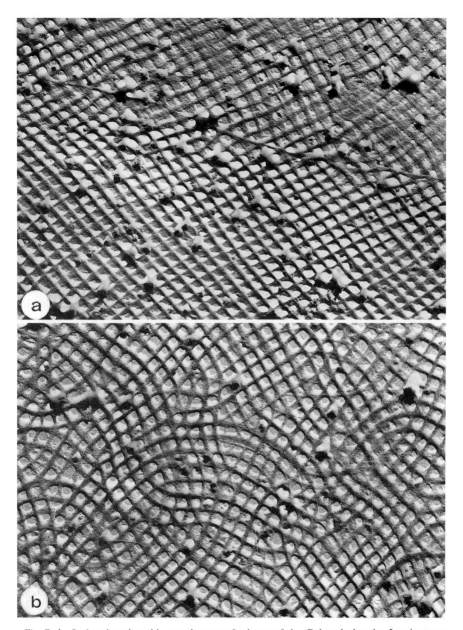

Fig. 7a,b. Isolated peritrophic membranes of a larva of the Colorado beetle, *Leptinotarsa decemlineata* (Coleoptera) after shadow-casting with platinum. *a* Regular and *b* irregular patterns were found; 18 000x (After Peters 1969)

23

from *Pieris brassicae*. Waterhouse observed about 150 species from 39 families of Lepidoptera (Table 4). In most of these, chitin-containing peritrophic membranes are secreted by specialized cells at the anterior end of the midgut. Only in three families (Hesperiidae, Satyridae and Nymphalidae) most of the midgut cells seem to secrete numerous fragile peritrophic membranes which surround the food in the gut. Moreover, Table 4 shows in which families peritrophic membranes are secreted by specialized cells at the entrance of the midgut (Type II after Wigglesworth 1930; Type I after Waterhouse 1953a). In Fig. 21 the secretion of peritrophic membranes and the formation of the texture of microfibrils in the larvae of the small tortoise shell, *Aglais urticae*, are shown.

Diptera: There seem to be only a few adult Diptera which have no peritrophic membranes (Fig. 8). This is not surprising in *Oestrus ovis* since the adults have no functional mouthparts and cannot take up food. A peritrophic membrane is also absent in the adult sheep ked, *Melophagus ovinus*, which sucks blood (Molyneux 1975; Molyneux et al. 1978). But it has been found in other species of the Hippoboscidae already by Aschner (1931), and in *Otholfersia macleayi* and *Ornithomyia* sp. by Waterhouse (1953a). A species of the Nycteribiidae, *Nycterebosca falcozi*, also has a peritrophic membrane of Type II. A peculiar exception in the Syrphidae is *Microdon modestus* in which Waterhouse could not find a peritrophic membrane.

Table 4. Occurrence of peritrophic membranes of Type II in adult Lepidoptera (After Waterhouse 1953a)[a]

Systematic group	PM		Systematic group		PM
Homoneura			Hypsidae	(1)	–
Hepialidea			Arctiidae	(4)	–
Hepialidae	(3)	–	Sphingoidea		
Heteroneura			Sphingidae	(1)	–
Gelechiodidea			Geometroidea		
Oecophoridae	(10)	–	Oenochromidae	(7)	–
Gelechiidae	(2)	+	Oenochromidae	(1)	+
Xyloryctidae	(1)	–	Boarmiidae	(2)	–
Xyloryctidae	(1)	+	Boarmiidae	(4)	–
Tortricoidea			Geometridae	(1)	+
Eucosmidae	(2)	+	Sterrhidae	(2)	–
Tortricidae	(6)	+	Larentiidae	(8)	+
Tineoidea			Bombycoidea		
Plutellidae	(1)	–	Saturniidae	(1)	–
Tineidae	(1)	–	Bombycidae	(1)	–
Pterophoroidea			Papilionoidea		
Pterophoridae	(3)	–	Danaidae	(2)	+
Pyraloidea			Papilionidae	(4)	+
Oxychirotidae	(1)	–	Pieridae	(6)	+
Phycitidae	(3)	–	Lycaenidae	(16)	+
Crambidae	(1)	–	Hesperiidae	(15)	–
Pyralidae	(1)	–	Satyridae	(11)	–
Pyraustidae	(7)	–	Nymphalidae	(4)	–

[a]The numbers in parentheses indicate the number of species of each family investigated.

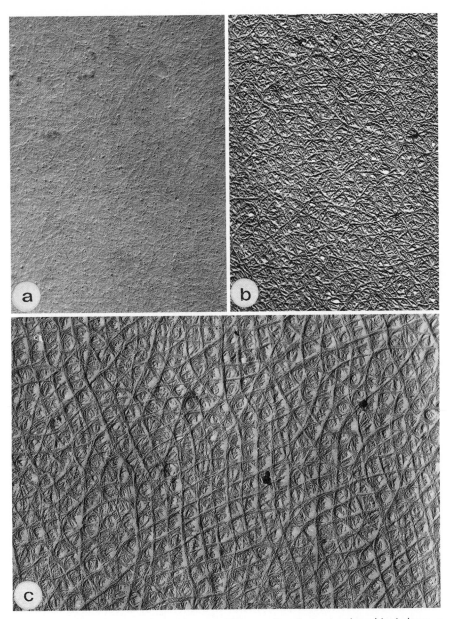

Fig. 8a-c. Isolated peritrophic membranes of Diptera after shadow-casting with platinum. *a* In the untreated inner peritrophic membrane of the adult blowfly, *Calliphora erythrocephala* (Diptera), the microfibrils are hidden by the matrix; 22 000x. *b* After prolonged treatment with 40% KOH for 20 h at 60 °C, a random texture of the microfibrils is evident; 30 000x. *c* In *Tipula* sp. larvae a regular pattern was found; 22 000x (After Peters 1969)

Blood-sucking species of four families secrete chitin-containing peritrophic membranes of Type II (according to Wigglesworth 1930) by delamination from larger parts of the midgut epithelium or from the whole midgut:

Culicidae: Aubertot (1932a,b), Waterhouse (1953a), Bertram and Bird (1961), Freyvogel and Jaquet (1965), Freyvogel and Stäubli (1965), Richardson and Romoser (1972), Romoser and Rothman (1973), Stohler (1957, 1961).

Phlebotomidae: Gemetchu (1974).

Simuliidae: Lewis (1950).

Tabanidae: Engel (1924), Aubertot (1934), Zhuzhikov (1963), Waterhouse (1953a).

Mutants

Despite the abundant knowledge which has been accumulated in genetics, structural mutants effecting the formation of peritrophic membranes have been described only exceptionally. In a strain of *Aedes aegypti*, the larvae produced such a large amount of peritrophic membrane that resistance to DDT resulted (Abedi and Brown 1961). The membranes served as a vehicle for DDT and DDE excretion. Rodriguez and Machado-Allison (1977) described in detail the genetics of a mutant called peritrofica in the same mosquito species. However, structural aspects were not mentioned except the fact that the peritrophic membrane protruded from the anus considerably, i.e. the length of the whole abdomen.

In *Drosophila melanogaster* a sex-linked lethal mutant strain 1(1)48J has been described in some detail by Rizki (1956). It is lethal during embryonic stages. The development of the midgut and muscles as well as other structures is disturbed. Sections of embryos 22–24 h old show an enormous accumulation of PAS-positive material in the cardia. The formation of the peritrophic membrane, if it is formed at all, is disturbed considerably. In such cases the membrane is unusually thick and irregularly shaped. Secretion products and peritrophic membrane are often sclerotized in late embryos. Rizki assumed that the malfunction can probably be explained by muscle deficiencies in the cardia.

14. Tentaculata

Phoronidea: Nothing can be said about peritrophic membranes in this small group.

Bryozoa: The occurrence of peritrophic membranes in Bryozoa is uncertain. If these animals are fed with small, degradable food organisms, such as bacteria and small flagellates, they seem to form no membranes around the fecal material. But if they are fed with flagellates, which have a thicker cellulose-containing cell wall or with ciliates like *Coleps*, the food residues are shed in the form of fecal pellets (Peters 1968).

Brachiopoda: *Lingula* sp. secretes peritrophic membranes no matter how much food is in the midgut. Microfibrils which are arranged at random are present in these membranes (Peters 1968).

15. Hemichordata

In *Glossobalanus* sp. and *Saccoglossus* sp. a delicate membrane which contained no chitin surrounded the food in the midgut and hindgut (Peters 1968).

16. Echinodermata

Antedon mediterranea seems to have no peritrophic membranes. *Cucumaria planci* surrounds the food residues with mucus which seems to contain no chitin.

In the gut of sea urchins, *Echinus esculentus*, *Psammechinus miliaris*, *Arbacia lixula* and *Araeosoma fenestratum* large amounts of fecal pellets were present in the gut. These were covered by a thin, translucent membrane. Even bulky material, e.g. the remains of crabs, was wrapped by such membranes, especially in *Araeosoma fenestratum*. Chitin or microfibrils could not be demonstrated in these membranes. It is remarkable that a sand-eating species like *Echinocardium cordatum* seems to form no peritrophic membranes (Peters 1968). The occurrence of mucous cells in the gut of 37 species of Echinoidea has been investigated comparatively by Holland and Ghiselin (1970).

17. Chaetognatha

Reeve et al. (1975) and Cosper and Reeve (1975) described for the first time peritrophic membranes in a chaetognath, *Sagitta hispida*. Hungry chaetognaths try to ingest one or more copepods within a few minutes and surround their remains eventually with a peritrophic membrane. The prey is thus transformed into elongate, irregular, fecal pellets. These are voided 3–4 h after ingestion. The peritrophic membrane is obviously a secretion product of the midgut. Peritrophic membranes without fecal material ("ghosts") were never seen.

18. Chordata/Tunicata

Appendicularia: *Oikopleura dioica* is able to secrete membranous material which surrounds the food residues in the gut (Fig. 9a). The regular texture of microfibrils depends on the arrangement of microvilli. A delicate, web-like pattern of chitin-containing microfibrils was found immediately after delamination from the microvilli of the midgut epithelium (Fig. 9b).

Thaliacea: *Salpa democratica* and *Thetys vagina* had long strands of fecal material (see also Fig. 7) in their guts which were surrounded by a transparent, delicate, but mechanically rather resistant membrane which did not dissolve in hot alkali. These peritrophic membranes gave a positive chitosan reaction only after prolonged hydrolysis for 7 days in 40% KOH at 60 °C. Probably the matrix material was highly resistant to such treatment. Electron microscopic investigations showed the presence of microfibrils which were arranged in a random texture but occasionally also in a hexagonal network (Peters 1968).

Ascidiacea: These filter feeders sample food particles in mucus secreted by the endostyle of the gill region. Then the mucus is wrapped into a strand, and transported to the midgut where the strand is surrounded by numerous peritrophic membranes. They are secreted by the midgut epithelium. Peters (1966, 1968) was able to demonstrate the presence of microfibrils arranged in a random texture (Fig. 10). He

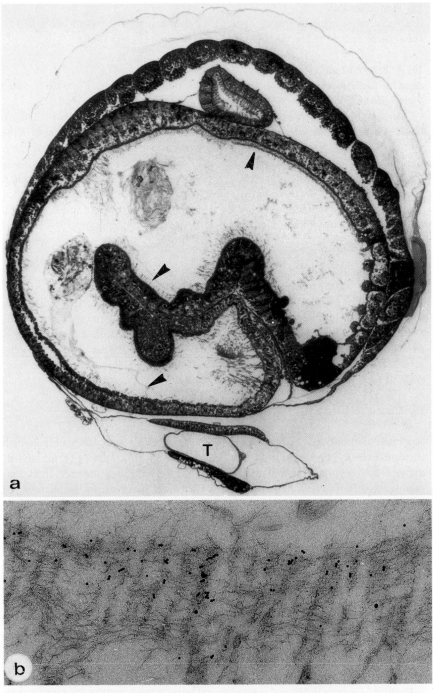

Fig. 9a,b. Oikopleura dioica (Tunicata, Appendicularia). *a* Light microscopic cross-section showing formation of peritrophic membranes in the midgut (*arrowheads*). *T* Tail; 52x. *b* Electron micrograph of a section through a peritrophic membrane in statu nascendi. The chitin content of the fine microfibrils is shown by specific labeling with wheat germ ag-glutinin-gold (see Sect. 5.2); 38 800x

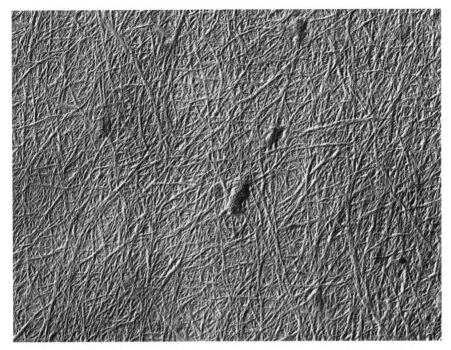

Fig. 10. Isolated peritrophic membranes of *Phallusia mammillata* (Tunicata Ascidiacea) after shadow-casting with platinum. They show a random texture of microfibrils; 35 000x (After Peters 1968)

also revealed the presence of chitin in the peritrophic membranes of the following species using the chitosan reaction: *Phallusia mammillata, Ciona intestinalis, Halocynthia papillosa, Clavelina lepadiformis* and *Corella* sp. In the small colony-forming species, *Sidnyum turbinatum* and *Botryllus schlosseri*, peritrophic membranes could be detected. The chitin content was demonstrated by the chitosan test. Rudall and Kenchington (1973) reinvestigated the problem. They found micro-fibrils in these membranes. After extraction with hot, dilute alkali, the residue did not show the chitin structure by X-ray diffraction nor by infrared absorption. However, the authors used preserved *Ciona intestinalis* and were obviously unable to differentiate between food residues and peritrophic membranes.

Therefore, the situation remains unsettled and should be reinvestigated again with appropriate modern methods. The problem is of considerable interest, since in the case of a positive result, the Tunicata could be the only group in the animal kingdom in which cellulose is formed in the mantle and chitin is secreted by the midgut epithelium. Hyman (1958), Florkin (1966) and Jeuniaux (1963) assumed an irreversible loss of the ability to synthesize chitin in the deuterostomia.

Acrania: *Branchiostoma lanceolatum* is able to secrete peritrophic membranes around the food residues which can be observed in the living animal. They occur from the caudal end of the peribranchial cavity to the anus. The peritrophic envelope rotates with considerable speed within the gut lumen, about one rotation per second. The peritrophic membranes around the fecal pellets remain intact in seawater for at least 7 days (Adam, cited in Peters 1967a). Nothing is known about the presence of chitin in these membranes.

Vertebrata

Agnatha: Adam (1960, 1963, 1965) described a peritrophic sac consisting of numerous peritrophic membranes in the gut of *Myxine glutinosa*.

Osteichthyes: Three species were investigated with respect to the occurrence of peritrophic membranes (Peters 1968). In *Gasterosteus aculeatus, Lebistes reticulatus* and *Xiphophorus helleri* the fecal strand may protrude considerably out of the anus before it disrupts and sinks to the ground. Fecal strands and pellets are held together by an envelope which consists of several membranes. There is no evidence of chitin in these envelopes.

Amphibia: *Triturus cristatus, Rana esculenta, Bufo viridis* and tadpoles of *Bufo bufo* showed the same features.

Reptilia: *Cordylus giganteus* and *Psammodromus hispanicus* have fecal balls which are covered with mucus (Peters 1968).

Aves: The feces of the nestlings of Passeres and Pici are surrounded by several layers of mucus. These envelopes are well known. They are secreted by the hindgut. The fecal balls may be relatively large and amount up to one-tenth of the nestling's weight. As soon as they are voided, the parents take them and eat them during the first days or remove them from the nest later on.

Mammalia: The feces of mammals may also be covered more or less by mucus. It has been assumed that the mucus in the gut of vertebrates is an analog to the peritrophic membranes. Berlese (1909) suggested already in his "*Gli Insetti*": "The peritrophic membrane roughly corresponds with the slime which surrounds food in the intestine of vertebrates." Glycoproteins of the surface coat and mucus are supposed to provide a barrier to microorganisms as well as enzymes and chemicals in the intestine (Spiro 1963; Ito 1965). Hollander (1954) assumed a two-component mucous barrier protecting the gastroduodenal mucosa against peptic ulceration. Forstner et al. (1973) were able to characterize for the first time a glycoprotein of high molecular weight representing the majority of the soluble glycoprotein hexosamine and hexose in the intestine. It could be localized in the goblet cells and the extracellular mucous coat. Antigenically similar material proved to be present in intestinal washings, stomach and colon. Some of the glycoproteins, which were specifically attached to microvilli, were shown to be disaccharidases (Forstner et al. 1973).

Chapter 3

Formation and Structure

3.1 Beginning of Secretion of Peritrophic Membranes

In most embryology papers nothing is said about the beginning of secretion of peritrophic membranes. Aubertot (1934) found these membranes already in immobile embryos in the brood pouch of water fleas, *Daphnia pulex*.

In embryos of *Drosophila melanogaster* the cardia is a well-formed organ after 12 h of development (Poulson 1950), but the secretion of the single peritrophic membrane does not start until after 18 h of development; hatching occurs after about 24 h of development.

In honeybees, *Apis mellifica*, Evenius (1925) and von Dehn (1933) observed a first and single peritrophic membrane in bees which were 19 days old, i.e. 19 days after the egg had been deposited. Twelve hours later three membranes were present, and 4 h after hatching still three or occasionally four membranes could be found (von Dehn 1933). Hering (1939) reported that the valvula cardiaca is not invaginated in bees which are 18 days old. About 1 day later it is fully formed and connects the lumina of foregut and midgut. According to Hering, a first peritrophic membrane appears in bees which are 20 days old. It extends through the whole midgut. The secretion of this membrane, and of those membranes which appear during the following days, takes place at the beginning of the midgut in several annular rows of cells. These cells have a dense plasma and stain more intensely than the remaining cells of the midgut (Loele 1914). The secretion of the peritrophic membranes starts already before the midgut opens into the hindgut (Rengel 1903; Hering 1939). Honeybees hatch 21–22 days after the egg has been deposited. Bamrick (1964) observed in larvae of the honeybee that peritrophic membranes appeared approximately 1.5 days after hatching. At the end of the feeding stage, the larva voids a peritrophic sac (see Sect. 3.3.2). The secretion of adult peritrophic membranes takes place already in the pupa. Two days before hatching flocculent material occurs in the midgut, and 4 days later peritrophic membranes can be isolated (Pabst et al. 1988).

In the mealworm beetle, *Tenebrio molitor*, Gerber (1976) observed that the lumen of the midgut of late pupae and young adults contains the meconium, the remains of the larval midgut. As soon as food is taken into the midgut, the remains of the meconium are transported to the hindgut and voided. The midgut epithelium is not fully functional until the second day after emergence. The formation of peritrophic membranes was observed in a small percentage of adults at day 1.5 and in most adults by day 2.5. At that time the epithelial cells change in appearance and the first adults start feeding.

In pupae of several species of mosquitoes, *Aedes triseriatus*, *A. sollicitans*, *Psorophora confinnis* and *Culex nigripalpus*, the presence of peritrophic membranes around the meconium was reported for the first time in holometabolous insects by

Romoser and Rothman (1973). A more detailed investigation by Romoser followed in 1974. The remains of the larval midgut, the meconium, are enveloped by a membrane which is secreted by the pupal midgut epithelium during the early phase of the pupal stage. Later in the pupal period a second membrane is secreted (Fig. 11). These membranes were termed peritrophic membranes for the following reasons: (1) they surround the contents of the midgut; (2) they appear to be secreted by the whole midgut epithelium in the same way as peritrophic membranes; (3) they give the same staining reactions; and (4) they reacted positively in the chitosan test. However, the latter may be true only for the second membrane which was detected first by Romoser and Rothman (1973). The remnants of these membranes form a mass in the blood bolus of blood-fed female mosquitoes. The meconial material is

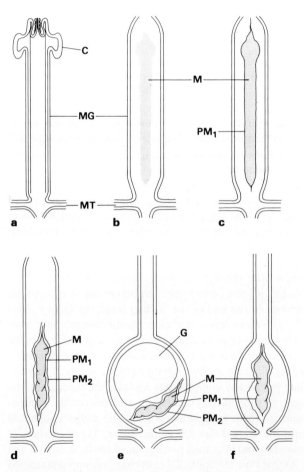

Fig. 11a-f. Formation of peritrophic membranes during pupation of mosquitoes shown in diagrammatic representations of sagittal midgut sections. a Larva; b recently emerged pupa; c young pupa; d old pupa; e recently emerged adult; f 10–24-h-old adult. C Cardia; G gas bubble; M meconium; MG midgut; MT Malpighian tubules; PM peritrophic membrane; PM 1 and 2 first and second peritrophic membrane formed by the pupa (Redrawn after Romoser 1974)

usually voided during the first 24 h after emergence of the adult mosquito. However, remnants of the membranes can persist even when the meconium has been voided as long as 33 days following adult emergence in females which had not been fed blood. Remnants of such peritrophic membranes, which were secreted during the pupal stage, were not only found in females but also in males of several species.

In flies the secretion of peritrophic membranes starts shortly after emergence. This was first reported for tsetse flies, *Glossina morsitans, G. palpalis* and *G. austeni,* by Willett (1966). Immediately after emergence no peritrophic membrane was found in the midgut. During the following 24 h, about 1 mm of peritrophic membrane was secreted per hour (Figs. 12, 13). It reached the hindgut after 3–4 days if the flies remained unfed. The same phenomenon can be observed in newly emerged blow-flies, *Calliphora erythrocephala* and *Protophormia terrae-novae*, where there are three continuous tube-like membranes, or in flesh-flies, *Sarcophaga barbata,* and houseflies, *Musca domestica*, with two peritrophic membranes.

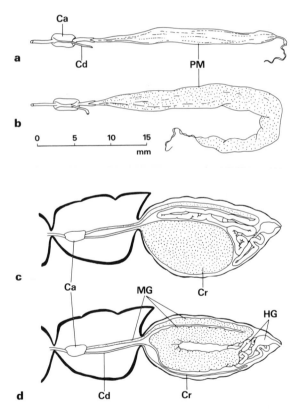

Fig. 12a-d. Formation of the peritrophic membrane (*PM*) and distribution of blood meal in the gut of *Glossina pallidipes. a* Peritrophic membrane of an unfed 32-h-old tsetse fly. On the *right* the beginning of formation with a twisted and fluid-tight end can be seen. *b* Peritrophic membrane of a 52-h-old fly, 30 min after feeding. *c* In a 24-h-old fly immediately after feeding, the crop is filled with the blood meal. *d* In a 40-h-old fly the blood meal has been pumped from the crop into the midgut. *Ca* Cardia; *Cd* crop duct; *Cr* crop; *HG* hindgut; *MG* midgut; *PM* peritrophic membrane (Redrawn after Harmsen 1973)

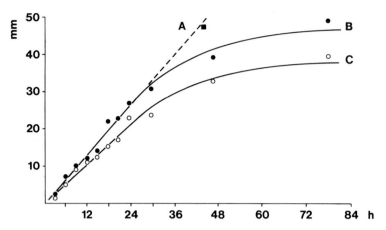

Fig. 13A-C. Growth of the peritrophic membrane in adult tsetse flies, *Glossina pallidipes*, after emergence. Length of membrane in mm is plotted versus age of fly in hours. *A* Flies fed 24 h at 25 °C. *B* Flies remained unfed (25 °C). *C* Flies remained unfed (30 °C) (After Harmsen 1973)

The posterior end of these peritrophic membranes is closed like a sac (Fig. 12). Approximately, the first 10 mm appear to be twisted and look like a long stocking or a sausage skin. The posterior end is not only twisted but tight. This can be proved by feeding suitable fluids such as blood, in the case of tsetse flies, or sugar solution labeled with a stain or fluorescent material. The statement of Wigglesworth (1929), based on sections of *Glossina tachinoides*, that in the newly emerged fly the membrane is ragged and discontinuous is not correct. The closed posterior end of the newly formed peritrophic membrane seemed to be of great importance in connection with the transmission of trypanosomes of the *Trypanosoma brucei* group (see Sect. 6.6.4).

3.2 Textures of Microfibrils

In ultrathin sections of peritrophic membranes of most species only microfibrils or aggregations of microfibrils can be seen in a matrix which does not stain after the usual preparation for electron microscopy (Figs. 14, 18, 21, 22, 29, 30). Exceptions are the highly differentiated membranes of Diptera with electron-dense layers; the latter can be stained owing to their proteoglycan content (Figs. 40–46). More information on the arrangement of these microfibrils can be obtained from whole mount preparations of single peritrophic membranes which are shadowed with platinum (see, for instance, Mercer and Day 1952; Peters 1968, 1969; Georgi 1969a,b) (Figs. 2–4, 6–8, 10, 16b, 24, 27). However, such preparations can only show the textures of microfibrils. Information on single microfibrils can only be obtained by preparations of single peritrophic membranes after negative staining

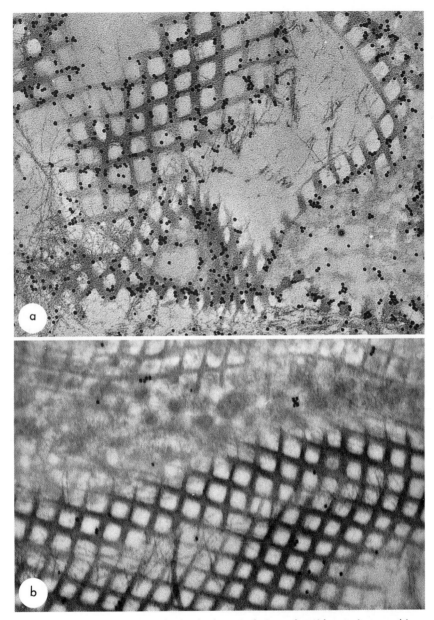

Fig. 14a. In the cardia of the earwig, *Forficula auricularia*, peritrophic membranes with an orthogonal texture of microfibril bundles are secreted which can be labeled with WGA-BSA-gold conjugate (for further details, see Sect. 5.2). It should be considered that despite the use of a water-soluble embedding material, Lowicryl K4M, only binding sites near the surface of the section can be labeled with lectin gold. Therefore, the majority of the binding sites in the interior of the section remain unlabeled; 36 000x. *b* After competition with 10 mM triacetyl chitotriose, the gold labeling on the microfibrils of these membranes has been almost completely abolished. Lowicryl K4M embedding; 37 000x (After Peters and Latka 1986)

with either phosphotungstic acid, uranyl acetate or ammonium molybdate (Rudall and Kenchington 1973; Streng 1973; Peters et al. 1979; Fig. 19).

Neville (1975) preferred the term crystallite rather than microfibril which is justified in terms of X-ray diffraction. In electron microscopy the term microfibril is used widely, and therefore will be retained here. The term microfiber is also used by some authors. Rudall and Kenchington (1973) determined the diameter of single microfibrils as 3 nm in negatively stained peritrophic membranes of drone larvae of the honeybee, and 6 nm in peritrophic membranes of weevil larvae, *Cionus scrophulariae*. Streng (1973) reported the same diameter for microfibrils in the peritrophic membrane of another weevil larva, *Rhynchaenus fagi*. In negatively stained peritrophic membranes of the earwig, *Forficula auricularia*, Peters et al. (1979) determined the diameter of single microfibrils to be in the range of 2.5 nm. In cuticles of arthropods single microfibrils had a diameter of 2.5 nm (Rudall and Kenchington 1973) or were in the range of 3 nm (Neville 1975).

Rudall and Kenchington (1973) observed in negatively stained cuticle preparations from the hornet a very distinct beading along the length of the microfibrils. The authors remarked that "there is good evidence for a periodic beading at ca. 3.1 nm, as expected from X-ray diffraction results". These surface structures might be due to protein as they disappeared when protein was removed enzymatically with pronase. The peritrophic membranes from a drone larvae of honeybees isolated with ultrasonics or preparations of single peritrophic membranes from the earwig, *Forficula auricularia*, did not reveal such beading after negative staining; they appeared as uniform strands. It was impossible to determine the whole length of a given microfibril. Peters et al. (1979) assumed that it exceeds 0.5 µm, as single microfibrils could be followed along that distance.

It has been generally assumed that these microfibrils contain chitin. They persist after protein has been removed and show in X-ray diffraction the characteristic diffraction pattern of chitin. Direct evidence was possible only recently in oblique sections of the inner peritrophic membranes of the earwig, *Forficula auricularia* (Peters and Latka 1986). If sections of material embedded in either Lowicryl K4M or LR White were incubated with colloidal gold labeled with the lectin wheat germ agglutinin, the gold granules were found predominantly on the orthogonal texture of bundles of microfibrils of the inner peritrophic envelope (Fig. 14). Specificity of binding was controlled by competition tests with 0.3 M N-acetylglucosamine and 5–10 mM triacetyl chitotriose.

Microfibrils may be arranged in three different types of textures (Fig. 1):

1. Random or felt-like texture.
2. Hexagonal or honeycomb texture.
3. Orthogonal or grid-like texture.

In most species the microfibrils are randomly distributed in the matrix of the peritrophic membranes. The regular hexagonal and orthogonal textures are restricted to peritrophic membranes. They have never been found in cuticles of animals or cell walls of plants. On the other hand, there seems to be no parallel arrangement of microfibrils in peritrophic membranes to form a lamellated system as in arthropod and other cuticles.

The formation of the regular textures of microfibrils has been a problem since they were first observed in preparations of single membranes shadowed with platinum. Mercer and Day postulated already in 1952 that the arrangement of microfibrils is determined or "templated" by the arrangement of the microvilli of cells which secrete chitin and matrix substances for the formation of peritrophic membranes: where the microvilli are staggered, a hexagonal texture of microfibrils should be formed, and where the microvilli are "in rank and file" an orthogonal texture should result (Fig. 23). Transitions were thought to be possible (Peters 1979).

This hypothesis has been favored by Georgi (1969b), Peters (1976), Peters et al. (1978, 1979), and Becker and Peters (1985) but criticized by Platzer-Schultz and Welsch (1969) and in more detail by Richards and Richards (1971, 1977); it has also been doubted by Rudall and Kenchington (1973).

Firstly, Platzer-Schultz and Welsch as well as Richards and Richards confused the regular patterns of chitin-containing microfibrils and proteoglycan-containing structures in the peritrophic membranes of larvae of Nematocera (Chironomidae, Culicidae and Simuliidae; Figs. 40–46). However, the proteoglycan-containing structures are of a different order of size as compared with the microfibrils. Their regular patterns appear to be formed by self-aggregation as there is no structure observable which fits their order of size.

Secondly, these authors regarded only species in which the secreted material for the formation of peritrophic membranes appears at the tips of microvilli. Therefore, they argued that the aggregation of microfibrils takes place some distance from the cell surface, without any possible influence of the pattern of microvilli. If in some micrographs the aggregation of peritrophic membranes seems to occur between the tips of microvilli, as has been stressed by Georgi (1969b), this was regarded by these authors as an artifact occurring during preparation, although Georgi could show a relationship between the diameter of microvilli and the diameter of the meshes of the microfibrillar network in several Decapoda (Crustacea).

The hypothesis of Mercer and Day could only be verified in species in which the polymerization of microfibrils occurs already in the interstices between the bases of extremely long microvilli, which excludes the argument of a possible displacement during preparation for electron microscopy. Some insect species fulfill these requirements (Peters et al. 1979).

The epithelium of the cardia of the earwig, *Forficula auricularia*, forms two annular folds, a deeper one and a shallower one (Fig. 15). In the first fold an inner peritrophic envelope consisting of several peritrophic membranes is secreted continuously. These membranes show an orthogonal texture of microfibril bundles which may have more or less disturbances (Fig. 16). In the second fold an outer peritrophic envelope consisting of numerous peritrophic membranes is secreted. In these membranes a random texture of microfibrils can be observed. In the midgut both types of envelopes surround food and food residues.

The cells of the first annular fold of the midgut epithelium (AFM 1 in Fig. 15) have extremely long microvilli with a length of 4.5–5.1 μm. Between the bases of these microvilli secretion products can be observed which are electron dense. Similar structures can be seen at different levels between the microvilli (Figs. 17, 18). In oblique sections they proved to be the orthogonal microfibrillar network of peritrophic membranes. Up to four successively formed membranes could be ob-

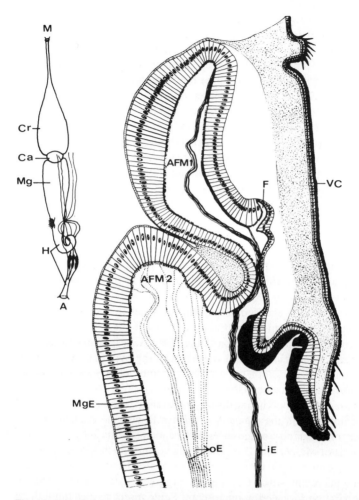

Fig. 15. Schematic representation of the cardia of *Forficula auricularia*. *Left* Alimentary canal: *A* anus, *Ca* cardia, *Cr* crop, *H* hindgut, *M* mouth, *Mg* midgut. *Right* Semidiagrammatic representation of a longitudinal section of the left side of the cardia and the adjacent part of the midgut. *AFM1* and *2* first and second annular fold of the midgut epithelium; *C* collar with thick cuticle; *F* furrow marking the end of the valvula epithelium; *iE* and *oE* inner and outer peritrophic envelopes, each consisting of several peritrophic membranes; *MgE* midgut epithelium; *VC* valvula cardiaca (After Peters et al. 1979)

→

Fig. 16. *a* The peritrophic envelope of *Forficula auricularia* consists of two types of membranes. The inner envelope *iE* is made up of more or less aggregated membranes with an orthogonal texture of microfibrils, whereas the outer envelope *oE* consists of numerous membranes with a random texture of microfibrils; 8000x. *b* Isolated peritrophic membrane from the inner envelope after shadow-casting with platinum. There are many irregularities in the orthogonal network; 18500 x (After Peters et al. 1979)

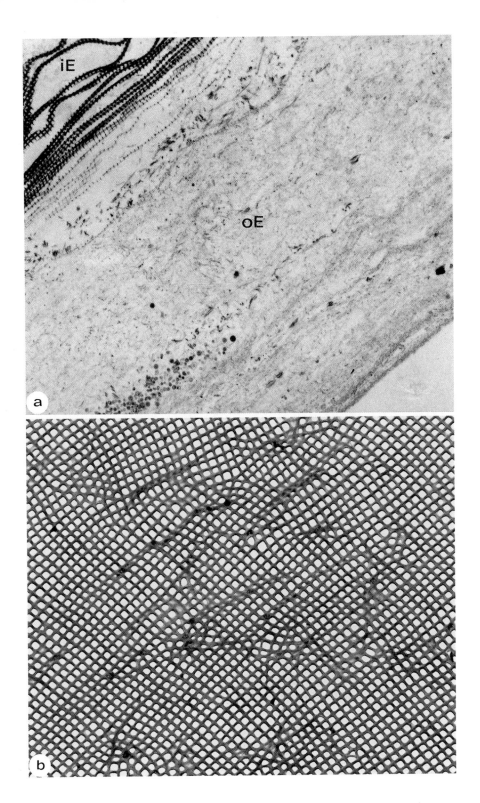

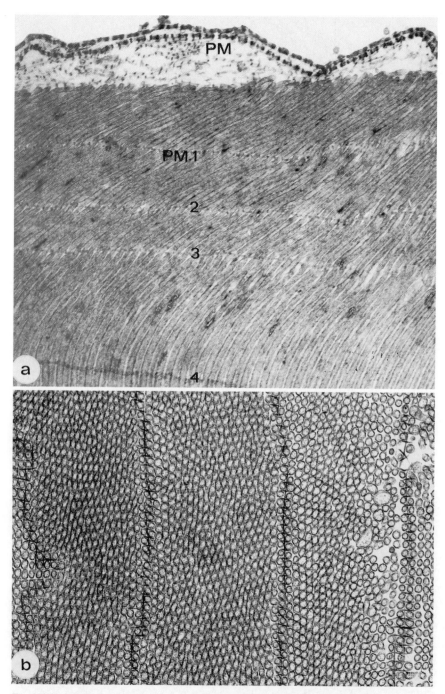

Fig. 17a,b. Forficula auricularia. a Up to four peritrophic membranes (*PM 1, 2, 3, 4*) were observed in the brush border of the formation zone in AFM1 and several peritrophic membranes with a different degree of aggregation on top of the microvilli; 2000 x. *b* In an oblique section of the brush border four newly formed peritrophic membranes are found; 16000 x (After Peters et al. 1979)

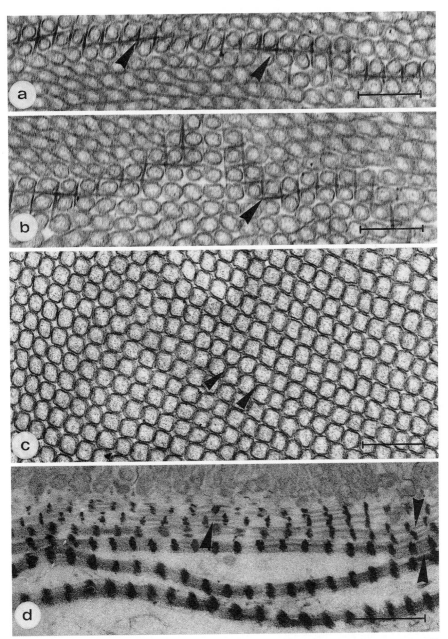

Fig. 18a-d. Forficula auricularia. a, b Newly formed bundles of microfibrils (*arrowheads*) between the highly ordered microvilli in these oblique sections through brush borders of the formation zone in AFM1. *c* Microvilli can change from a circular to a more square outline in transverse sections and from rank and file to a more staggered arrangement (*arrowheads*). Such disturbances are reflected in the texture of microfibrils (see Fig. 19). *d* Different degrees of aggregation of peritrophic membranes (*arrowheads*) are found on top of the very long microvilli of the formation zone of AFM1. The *dark points* mark the crossings of microfibril bundles. See Fig. 19e. Scale: 0.5 μm (After Peters et al. 1979)

served in the microvillous border (Fig. 17). In cross-sections of the brush border a very regular pattern of microvilli can be seen in larger areas. The microvilli are arranged in rank and file. Sometimes they are packed so closely that they appear rectangular instead of showing the usual round cross-section. In a few places the orthogonal pattern of microfibril bundles was found between the highly ordered microvilli. The diameter of the microvilli was in the range of 145 ± 12 nm, and the inside width of the orthogonal network of microfibril bundles was in the range of 125 ± 14 nm.

Single peritrophic membranes, which were isolated from the inner peritrophic envelope and negatively stained, contained bundles consisting of about ten microfibrils. The bundles had a diameter in the range of 20 nm and the diameter of a single microfibril was about 2.5 nm. In such negatively stained preparations there was no periodic beading of protein components detectable around the microfibrils (Fig. 19). At the crossings of the bundles there was not much felting of microfibrils which means that the orthogonal pattern is achieved by putting two layers of perpendicularly oriented microfibril bundles one upon another (Figs. 19, 20).

In sections of the formation zone as well as in negatively stained preparations, peritrophic membranes of different thickness were found. Obviously, several peritrophic membranes stick together soon after formation to form aggregated peritrophic membranes (Fig. 18). No indications of a common matrix or of sticky material could be observed in electron micrographs. As the crossings of the bundles of microfibrils are in register, these aggregated membranes are heavily contrasted by the usual staining techniques for electron microscopy. Therefore, they are more easily detected in the brush border of the formation zone and show more clearly the orthogonal pattern than single membranes. As the microfibrils are not exactly in register in these aggregated membranes, there is a reduction of the inner width of the grid squares down to 60–80 nm. Sometimes the regular pattern of microvilli is disturbed (Fig. 16b). Single microvilli may be absent, disorders may occur along cell borders, or groups of microvilli may have a staggered arrangement. The latter results in a hexagonal texture of microfibrils. The alterations may concern single microfibrils, groups of microfibrils or even whole bundles (Fig. 20). If microfibrils leave their bundle they may change their track or their direction.

In the formation zone a peritrophic membrane can be traced through the brush border of several cells and then projects more and more out of it. Finally, about 20 peritrophic membranes can be found on top of the microvilli where they lie parallel to each other or in irregular folds (Figs. 17, 18). There are aggregated as well as single peritrophic membranes, and the latter can be found also in peritrophic envelopes from the midgut (Fig. 16).

Wigglesworth (1930) assumed that in *Forficula* the material for the formation of peritrophic membranes is secreted as a viscous fluid which is molded into a thin tube by an annular thickening of cuticle at the end of the valvula cardiaca, the so-called cardia press (Fig. 47). However, this concept was already rejected by Giles (1965) due to light microsopic investigations of another earwig, *Anisolabis litorea*. Only a few cells are involved in the formation of the inner peritrophic membranes of *Forficula auricularia*. Nobody knows the function of the similarly appearing cells of the adjoining epithelium.

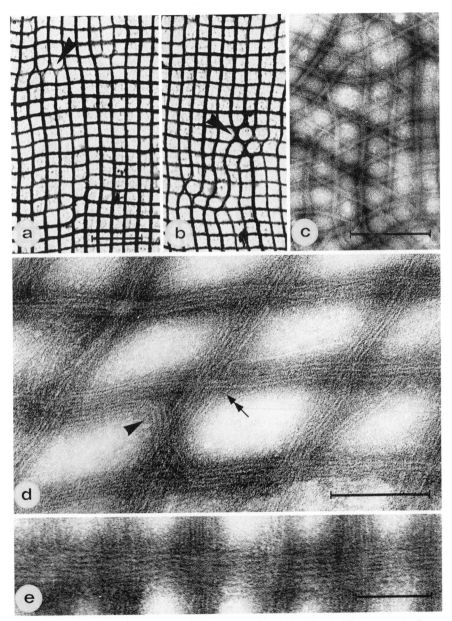

Fig. 19a-e. Forficula auricularia. Negatively stained peritrophic membranes of the inner peritrophic envelope. *a-c* The staggered arrangement of microvilli leads to irregularities of the orthogonal texture of microfibril bundles (*arrowheads*) and can even cause a hexagonal lattice of the bundles (*b,c*) The *scale lines* are equivalent to 0.5 µm. *d* Sometimes irregularities of single microfibrils changing track (*arrowhead*) or felting at the cross-points of bundles (*double arrow*) can be observed. The *scale line* is equivalent to 0.1 µm. *e* In the orthogonal lattice the parallel microfibril bundles of one direction often lie upon the bundles of the perpendicular direction. Scale: 0.1 µm (After Peters et al. 1979)

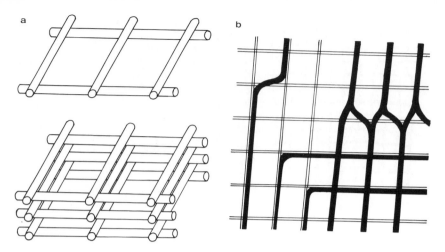

Fig. 20a,b. Forficula auricularia. a Two stages of the aggregation of single peritrophic membranes of the inner peritrophic envelope. b A summary of the main irregularities found in the texture of the inner peritrophic envelope (from *left to right*): microfibrils changing track, changing direction, and changing bundles caused by staggered microvilli (After Peters et al. 1979)

Giles (1965) pointed out already that there is still another problem. How can a continuous, tube-like peritrophic membrane be formed in these earwigs? He suggested that the formation of these tubes might be achieved in the same way as the ribbon of paraffin sections on a microtome: After a length of membrane is formed it is passed backwards and joined to the next and so on. One should expect therefore periodic, seam-like structures. However, these have never been found. Those which were tentatively assumed by Giles (his Fig. 14) were probably only irregular distortions. Therefore, it seems to be more reasonable to assume that the inner peritrophic envelope of the earwig is not a continuous tube but consists of an aggregation of patch-like peritrophic membranes.

The verification of the hypothesis of Mercer and Day (1952) is not only restricted to *Forficula auricularia*. The template function of the microvilli could also be shown in the larvae of a butterfly, the small tortoiseshell, *Aglais urticae* (Fig. 21), in the larvae of Ephemeroptera (Fig. 22), and in a harvestman, *Phalangium opilio* (Becker and Peters 1985; Fig. 29c). In the larvae of *Aglais urticae* the microvilli of the midgut cells which secrete material for peritrophic membranes are extremely long. Again, several peritrophic membranes can be found simultaneously in the microvillous border. In this case the texture is hexagonal. The microvilli of the formation cells in larvae of Ephemeroptera are not as long as those of the preceding species. Nevertheless, the polymerization of the chitin-containing microfibrils takes place in the interstices between the bases of microvilli.

Georgi (1969b) pointed out that the relation between the area of the cross-sections of microvilli and the area of the interstices between microvilli or the total area is of importance for the texture of microfibrils. In other words, if the number of microvilli decreases in a given area, the number of microfibrils or bundles of

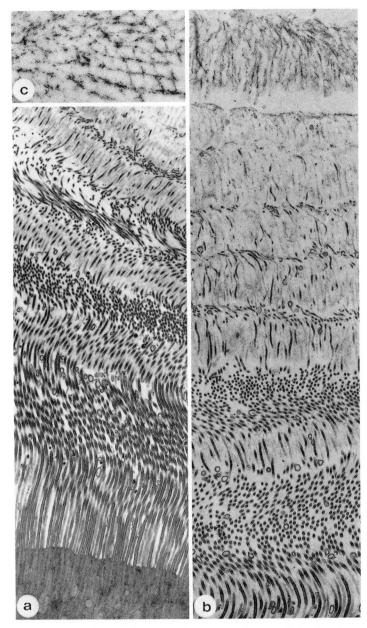

Fig. 21a-c. Formation of peritrophic membranes in the extremely high brush border of the midgut of a butterfly larva, the small tortoiseshell, *Aglais urticae. a* Survey; 5000x. *b* Detachment of peritrophic membranes; 7000 x. *c* Regular texture of microfibrils;   15700 x

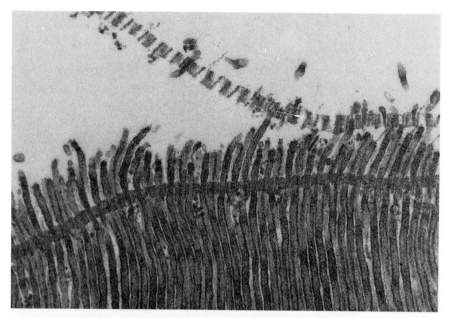

Fig. 22. Formation of peritrophic membranes in the brush border of the midgut of a mayfly larva, Ephemeroptera. One of these membranes has just detached, whereas the next one is already fully formed but still situated between the upper parts of the very long microvilli. This is another example of the template function of microvilli during the formation of peritrophic membranes; 30 000x

microfibrils might increase (Fig. 23). However, such patterns have not been observed as yet.

Another mechanism for the formation of regular textures of microfibrils seems to exist in larvae of Nematocera, in which a single, continuous, tube-like peritrophic membrane is secreted by a few rings of specialized cells in the cardia (Peters 1979). Blackfly larvae, *Odagmia ornata* (Simuliidae), are able to form the most complicated regular structures which have been found in peritrophic membranes until now (Fig. 46). At first a thin layer of finely granular material appears on top of the microvilli. As the secreted material thickens, two regular patterns can be discerned. These seem to be formed by self-aggregation from the amorphous material. On the lumen side a layer of small electron-dense rods with a length of about 30 nm can be seen. Four of these small rods are attached to the edges of compact rods of a second layer. This relation is clearly revealed in tangential sections (Fig. 46d) where the elements are cut at different angles. In some places the strictly orthogonal pattern may be disturbed (Fig. 46d and e). In cross-sections the two elements look like tuning forks. The compact rods have a width of about 20 nm and a length of about 30 nm. The electron-lucent spacings between the compact rods are in the range of 10 nm, those in a "tuning fork" are about 20 nm, and those between two tuning forks are about 10 nm, the same as between the compact rods. Histochemical reactions proved that the rods of both layers contain proteoglycans. After the development of

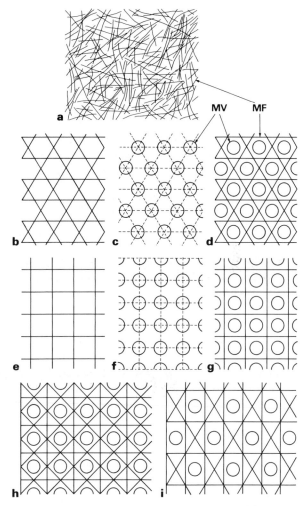

Fig. 23a-i. A schematic representation of the possible arrangements of microvilli and textures of microfibrils. Only three arrangements are realized (see Fig. 1). *a* Random texture. *b-d* Hexagonal texture with three directions: arrangement of microfibrils (*b*), microvilli (*c*), microvilli and microfibrils (*d*). *e-g* Orthogonal texture with two directions: pattern of microfibrils (*e*), microvilli (*f*), microvilli (*MV*) and microfibrils (*MF*) (*g*). *h-i* Different relations between area of cross-section of microvilli to total area are not realized (After Georgi 1969b)

these two layers additional material is secreted in which an orthogonal texture of the bundles of microfibrils appears. The spacings are in the range of 110–130 nm if measured from the middle of a bundle to the next. This is in good agreement with the width of the microvilli which is in the range of 85–95 nm.

The grid texture of these bundles of microfibrils is arranged in layers which are oriented parallel to the surface of the epithelium of the formation zone in the cardia; five to seven layers seem to be present usually. Finally, the membrane has a thickness of about 2 µm. Although the spacings between the bundles of microfibrils are in the same range as the width of the microvilli, there cannot be a direct influence on pattern formation as in the preceding examples; firstly, because the formation of the peritrophic membrane seems to occur on top of the microvilli, and secondly, because the two layers of rods separate the layer containing the bundles of chitin-containing microfibrils from the tips of the microvilli. Furthermore, this example shows that microfibrils polymerize outside the range of the microvilli and cannot be directed by the cell. Richards and Richards (1977) found similar elements of three different magnitudes (approximately 15, 35 and 120 nm) in *Simulium venustum*. In terms of center to center spacings there is a relationship of 1:2:8. It is tempting to assume that these relationships might influence the formation of the orthogonal pattern of microfibrils.

In the larva of the yellow fever mosquito, *Aedes aegypti*, there is a thick, moderately electron-dense layer of homogeneous appearance (Fig. 44a-d), instead of the layers of rods, covering the main part of the membrane with its orthogonal network of bundles of microfibrils (Zhuzhikov 1969; Richards and Richards 1977; Peters 1979). But perhaps there may be a periodic but invisible arrangement of compounds which direct the regular arrangement of the bundles of microfibrils in the following layer.

It has been mentioned already that flaws occur in the textures of microfibrils (Figs. 19, 20) as well as disorder in the arrangement of microvilli (Figs. 4, 6, 7, 16). Sometimes these are abundant as shown in Fig. 7. In terms of mathematics this means the compensation of two groups of curves which cross each other. This has been analyzed mathematically by Rosin (1946).

Transitions from a staggered arrangement of microvilli into an arrangement "in rank and file" occur as well as transitions from a hexagonal to an orthogonal pattern of microvilli. Rajulu (1971) investigated shadow-casted preparations of peritrophic membranes from different parts of the midgut of a chilopod, *Ethmostigmus spinosus*. He found a regular orthogonal pattern of microfibrils. The grid spacings changed from 23–38 nm in the anterior part, to 36–49 nm in the middle part, and to 46–58 nm in the posterior part. Unfortunately, Rajulu did not regard the width of the microvilli in the respective parts of the midgut.

Comparative investigations of the texture of microfibrils in the peritrophic membranes of 75 species of Insecta from 16 orders (Peters 1969) and of 43 species of Crustacea from 14 orders (Georgi 1969) revealed the following:

1. In random textures the degree of strand formation is very different. Rather thick strands were observed for instance in the peritrophic membranes of bumblebees, *Bombus lapidarius* (Fig. 24c). Apparently, there is no relation between the texture of microfibrils and the type of food, as the honeybee (Fig. 24a), which uses the same type of food as the bumblebee, has very thin microfibrils like the wasp, *Vespa germanica* (Fig. 24b), which prefers a variety of food.
2. The hexagonal texture is apparently always a reinforcement of the random texture. It occurs only on one side, i.e. on the epithelium side of a random texture; this was confirmed by Schlecht (1979). The diameter of a hexagon, measured as the distance between the main strands, was determined in shadow-cast preparations of peritrophic membranes from ten species of Decapoda

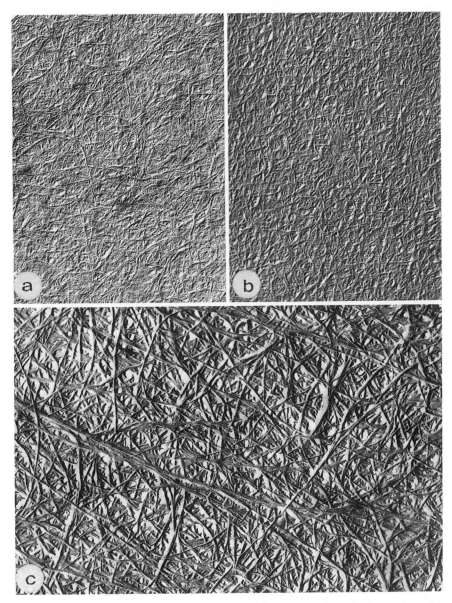

Fig. 24a-c. Isolated peritrophic membranes of Hymenoptera after shadow-casting with platinum reveal a random texture of microfibrils; 24 000x. *a* Honeybee worker, *Apis mellifica*; *b* wasp worker, *Vespula germanica*; *c* female of the bumblebee, *Bombus lapidarius* with microfibril bundles (After Peters 1969)

(Crustacea) by Georgi (1969). It proved to be 130–260 ± 10–30 nm. The orthogonal texture seems to occur very often without a random texture.

3. In a given species and its stages the same type of texture or combination of textures was observed.

4. In Crustacea the hexagonal texture was found by Georgi only in Malacostraca, but Schlecht (1979) reported it also from Phyllopoda, *Daphnia pulex* and *Leptestheria dahalacensis*, whereas the orthogonal texture occurred only in Anostraca and Notostraca.

5. A phylogenetic trend beginning with a random texture in primitive forms and ending with an orthogonal texture in more advanced forms does not exist.

6. There seems to be no simple correlation between type of nutrition and texture of microfibrils in peritrophic membranes. But Georgi (1969) suggested already that the reinforcement of a random texture by a hexagonal network of microfibrils might require the availability of food and therefore of material for the synthesis of peritrophic membranes. In other words, a random texture is always formed, but hexagonal patterns occur only if a certain amount of food is available. This was confirmed by Schlecht (1979) in *Leptestheria dahalacensis* (Phyllopoda). Well-nourished animals secreted peritrophic membranes with a random, hexagonal and orthogonal texture of microfibrils. But hungry animals with an empty gut could apparently form neither ordered patterns of microfibrils nor regular peritrophic membranes. In such animals the midgut was filled by an irregular mass of microfibrils.

3.3 Formation of Peritrophic Membranes

In connection with the old postulate that only cells of ectodermal origin are able to synthesize and secrete chitin, there has been a controversy about the question whether peritrophic membranes are secreted by cells of endodermal or ectodermal origin. It has always been regretted that this problem cannot be settled in insects since in this group the endodermal origin, at least of parts of the midgut, is still problematic. However, if the insects are disregarded, there are good examples in other groups of animals where there is no doubt that the midgut originates from the endodermis. Such an example is the earthworm, *Lumbricus terrestris*. Menzi (1919) could show that the highly differentiated anterior parts of the gut, up to the end of the pharynx, are of ectodermal, whereas the crop, gizzard and midgut are of endodermal origin. The midgut epithelium is able to secrete peritrophic membranes which contain chitin microfibrils (Fig. 25). The gizzard secretes continuously material which contains chitin and enzymes, among other substances (Peters and Walldorf 1986a,b). This material was called cuticle, but it seems more appropriate to call it a gastric shield as in Mollusca where a similar structure occurs in all groups (Owen 1966). It is abraded on the lumen side by food particles and completely renewed after 48–60 h, as could be shown by radioactive labeling. Biochemical investigations proved that chitin is synthesized continuously by the gizzard epithelium. So there is no doubt that in this case two different chitin-containing materials

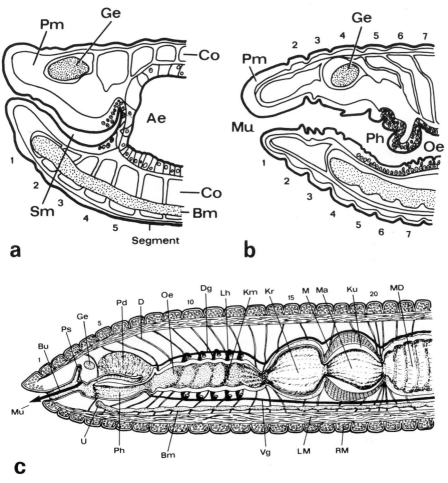

Fig. 25a,b. Schematic representation of the development of the foregut of the earthworm, *Lumbricus terrestris* (Annelida, Oligochaeta), showing the different parts of the alimentary tract. *Ae* archenteron; *Bm* ventral nerve cord; *Co* coelomic cavity; *D* dissepiment; *Ge* brain; *Mu* mouth; *Oe* esophagus; *Ph* pharynx; *Pm* prostomium; *Sm* stomodaeum. The *numbers 1–5* represent the number of the respective segment (After Menzi 1919 from Peters and Walldorf 1986b). *c* Diagrammatic representation of a median section through the anterior part of the adult earthworm, *Lumbricus terrestris*, showing the differentiation of the gut (genital organs omitted). *Bu* buccal cavity; *Bm* ventral nerve cord; *D* dissepiment; *Dg* dorsal blood vessel; *Ge* cerebral ganglion; *Lh* lateral heart; *LM* longitudinal muscles; *Km* opening of the calciferous glands; *Kr* crop; *Ku* cuticle of the gizzard containing enzymes and chitin; *Mu* mouth through which the pharynx pad (*Ps*) can be extruded; *MD* midgut; *M* muscle; *Oe* esophagus; *Pd* pharynx gland; *RM* circular muscle; *U* subpharyngeal ganglion; *Vg* ventral blood vessel; *1–20* number of segment (After Peters and Walldorf 1986b)

are secreted by cells which are evidently of endodermal origin. Those parts of the foregut which are of ectodermal origin, such as the buccal cavity and pharynx, do not secrete chitin but a collagen-containing cuticle such as the epidermis.

3.3.1 Formation Zones

In 1930 Wigglesworth classified peritrophic membranes according to their formation zones.

Type I peritrophic membranes are formed by the whole midgut epithelium. In the light microscope it appeared that patch-like membranes are delaminated from smaller parts of the epithelium, and that these patches are united to form a peritrophic membrane. Electron microscopy proved that the resolution of the light microscope is not sufficient to detect that the patches are single membranes which together form a peritrophic envelope consisting of several membranes.

Type II is characterized by a restriction of the ability to form peritrophic membranes by a belt of cells at the entrance of the midgut which is only several rows of cells wide; flies and larvae of mosquitoes are well-known examples.

Unfortunately, Waterhouse (1953a) used an inverse definition of Types I and II. Generally, the definition given by Wigglesworth (1930) is used, as it appears more logical to regard the formation of peritrophic membranes by the whole midgut epithelium as more ancestral than the formation by a restricted zone at the entrance of the midgut. For many years there was a controversy as to which of the two types was present in a given species, if not both, as for instance in bees (Schreiber 1956). Obviously, this question is as useless as the question, which was first, the hen or the egg? It is a typical man-made problem due to a premature classification. Electron microscopic investigations have shown that there is no reason for a distinction between delamination in Type I and secretion in Type II as in both cases secretion takes place (Richards and Richards 1977).

In the following it is not intended, therefore, to increase the number of types of formation from two to six; it is just an enumeration of combinations to obtain a better survey. Over more than 50 years after Wigglesworth described two different types of formation, the following variations of formation of peritrophic membranes have been found:

1. Originally, and still in most animal phyla, the whole midgut is able to secrete peritrophic membranes.
2. In several groups, especially of insects, the ability to form peritrophic membranes is restricted either to the anterior part of the midgut, or
3. to the posterior part of the midgut, or
4. it is restricted to annular groups of cells at the entrance of the midgut, or to
5. annular groups at the end of the midgut, or to
6. annular groups of cells in the middle third of the midgut.

3.3.2 Formation by the Whole Midgut Epithelium

There may be considerable differences in the ability to secrete peritrophic membranes and in the amount of secretory products shed in different areas of the midgut, in cross-sections as well as along the length of the midgut. Firstly, there is no synchronization of the whole midgut epithelium to secrete material for peritrophic membranes. Only smaller areas appear to be synchronized, since in the light microscope small patches of membrane can be observed leaving the brush border of the midgut epithelial cells. This has been called delamination (Figs. 9a, 26a). In electron micrographs secreted material appears as a more or less electron-dense fuzzy or more compact aggregation either between or on top of the microvilli (Figs. 9b, 17, 21, 22).

After the patch-like peritrophic membranes are detached from the microvillous border of the midgut cells by an unknown mechanism, they are used to surround food and food residues in the lumen of the midgut. It is, however, erroneous that they are used only as an envelope. Obviously, it depends on the time of formation and on the consistence of the incoming food as to how much the peritrophic membranes appear to be mingled with the contents of the midgut (Figs. 26, 29). A few examples may illustrate this.

In the earthworm, *Lumbricus terrestris*, nearly the whole midgut is able to secrete peritrophic membranes (Vierhaus 1971). This is less pronounced in the anterior part (segments 20–30), where secretion takes place mainly in the dorsolateral region. Nevertheless, the lumen of this part of the midgut is usually crowded with peritrophic membranes. The maximum membrane formation was found in the segments 30–100 in worms which had about 130 segments. This means that the midgut down to the end of the typhlosolis is able to secrete peritrophic membranes. In cross-sections of different regions Vierhaus saw that the epithelium of the typhlosolis, perhaps with the exception of its lateral parts, secretes less membrane material than the remaining parts. Light microscopy of longitudinal sections revealed that developing patches which were still in contact with the epithelium represent single membranes. These may extend for more than 10–15 segments. Combined with the results obtained with cross-sections, this shows that single membranes may have a surface area of about 80 mm^2. Earthworms have a marked diurnal rhythm of food intake; they feed mainly between 19:00 and 24:00 h in the evening. Although there is a more or less continuous production even in hungry worms, Vierhaus observed an increased formation of membranes at night, between 01:00 and 05:00 h.

In the Crustacea Decapoda the midgut and its diverticula, the caeca and the midgut gland contribute to the formation of peritrophic membranes. Georgi (1969b) described in the freshwater species *Orconectes limosus* (*Cambarus affinis*) that peritrophic membranes are secreted by the midgut epithelium, the dorsal caecum and by the duct of the midgut gland. He did not mention that peritrophic membranes are also formed in the tubules of the midgut gland.

This was reported for the first time in *Carcinus maenas* by Hopkin and Nott (1980; Fig. 28). In the midgut gland all epithelial cells secrete peritrophic membranes. These appear on top of the microvillous border as a continuous sheet which at last detaches from it. Several peritrophic membranes may be present simultane-

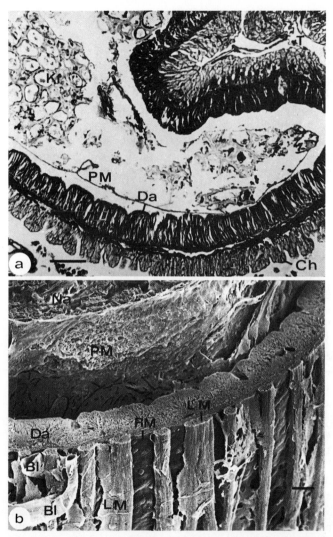

Fig. 26a. Transverse section of the midgut of the earthworm *Lumbricus terrestris*. Peritrophic membranes (*PM*), are delaminated, which surround the remnants of food, in this case coffee grounds (*Kf*). *Ch* Chloragog; *Da* midgut epithelium; *T* typhlosolis. Scale 100 μm. *b* Scanning electron micrograph of a transverse section through the midgut of the earthworm without chloragog. It shows the peritrophic membranes (*PM*) forming a sheath around the remnants of food (*Na*). *Bl* Bloodvessel; *LM* longitudinal muscles; *RM* circular muscles. Scale 10 μm (After Peters and Walldorf 1986b)

Fig. 27. A random texture of microfibrils occurs in the peritrophic membranes of the earthworm, *Lumbricus terrestris.* Isolated membranes after shadow-casting with platinum; 38 000x (After Peters 1968)

ously in the lumina of the ducts of the midgut gland. They enclose cellular debris, sulfurous bodies and B-cells extruded from the epithelium. During the second stage of the digestive cycle, an efficient mechanism provides the removal of the peritrophic membranes with the material enclosed by them and the transfer to the midgut. However, there is always a peritrophic membrane which remains and covers the surface of the epithelium. The epithelium at the anterior end of the midgut secretes a clear, membranous tube, consisting of several peritrophic membranes, which envelops the column of material in the lumen. One to 24 h after feeding, the peritrophic envelope contains the coarse residues of primary digestion. Between approximately 24–48 h after feeding the material in the midgut gland passes into the midgut. When the valve between the midgut and hindgut opens, the contents of the midgut pass into the hindgut and form the fecal column. The latter is voided in three portions. The first portion is voided about 8 h after feeding. It contains the remains of the previous meal. A second portion is voided 24–36 h after feeding and consists of the coarse particles which are too large to enter the lumina of the ducts of the midgut gland. A third portion is partially voided 48 h after the next meal. It contains the same type of material wrapped in peritrophic membranes as the first portion.

In the Decapoda Anomura and Brachyura usually two types of midgut caeca exist: a posterior dorsal caecum which enters the midgut at its junction with the hindgut and a pair of anterior caeca which join the midgut just posterior to the pylorus. According to Holliday et al. (1980), these caeca do not have a significant function in the formation of peritrophic membranes or in the osmoregulation in the crab *Cancer magister.*

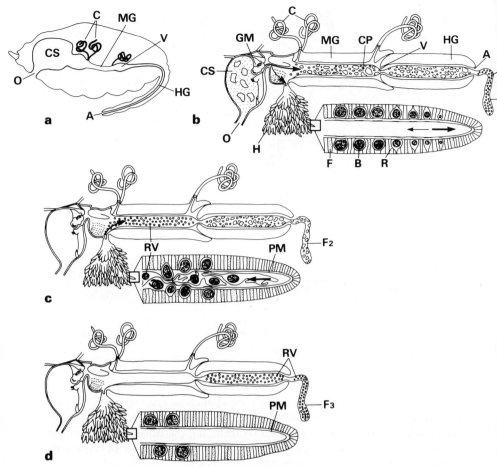

Fig. 28a-d. Three stages of the digestive cycle of the shore crab, *Carcinus maenas*. Schematic diagrams, not to scale. *a* General survey. *A* anus; *C* caeca; *CS* cardiac stomach; *HG* hindgut; *MG* midgut; *O* mouth; *V* valve. *b* 0–12 h after feeding: digestive and absorptive stage. Food fragments in the cardiac stomach (*CS*) are ground by the gastric mill (*GM*) and sorted by setae in the pyloric stomach into a coarse component (*CP*), which passes into the midgut (*MG*) (*upper arrow*), and a component consisting of liquid and fine particles which passes into the midgut gland or hepatopancreas (*H, lower arrow*). In the lumen of the tubules of the midgut gland, this material is moved back and forth. Some of the F cells (*F*) in the midregion of the tubules develop into B cells (*B*) in which digestion of the soluble components which pass through the peritrophic membranes is thought to take place. The first section of the fecal column (*F1*) is voided. It consists of residual vacuoles (*RV*) and peritrophic membranes derived from the midgut gland and held in the hindgut (*HG*) since the end of the previous digestive cycle. *A* Anus; *C* caecum; *O* esophagus; *R* assimilation and storage R cell; *RV* residual vacuole; *V* valve. *c* 12–48 h after feeding: excretory stage. Digestion is completed and numerous B cells are extruded into the lumen of the midgut gland. The residual vacuoles (*RV*) of these cells and the peritrophic membranes (*PM*) are passed into the midgut (*arrows*). The second section of the fecal column (*F2*) is voided. It consists of the coarse residues of primary digestion which are too large to enter the ducts of the midgut gland. *d* 48 h or more after feeding. The third section of the fecal column (*F3*) is voided which consists of residual vacuoles (*RV*) and peritrophic membranes (*PM*) derived from the midgut gland. The end of this section remains in the hindgut until the next meal (After Hopkin and Nott 1980)

The widely accepted concepts of food movement in Decapoda do not apply to the gut of larval and early postlarval stages of the white shrimp, *Penaeus setiferus*. Video microscopy in vivo has shown that the foregut does not masticate food but conveys only the food particles which were masticated by the mandibles into the midgut. In early postlarvae food is transported directly into the peritrophic membranes of the abdominal midgut. Later on it is regurgitated and moved by peristalsis into the hepatopancreas. This is facilitated by the uptake of water through the anus.

Different types of peritrophic membranes are secreted and different digestive functions occur in the two parts of the midgut of harvestmen (Opilionida) (Peters 1967b; Becker and Peters 1985). The large midgut is divided into an anterior and a posterior compartment by a dorsal fold (Fig. 29). The spacious anterior part of the midgut and the midgut gland together form a functional unit; they digest and resorb nutrients. The indigestable food residues and excreta pass into the posterior part of the midgut where they are compressed, dehydrated and surrounded by numerous peritrophic membranes to form a large fecal pellet. The epithelium of the anterior midgut secretes peritrophic membranes in a period between 2–4 h after feeding. Chitin-containing microfibrils polymerize between the basal parts of the microvilli and form a hexagonal texture. Irregularities of this pattern can be correlated with those of the microvilli. Thick peritrophic membranes (300–500 nm) with a hexagonal texture of microfibrils are accompanied by thin membranes (100–300 nm) with a random texture of microfibrils. A maximum number of up to four layers of thick peritrophic membranes was observed surrounding the mass of nutrients. In contrast to the anterior part, the microvillous border of the epithelium of the posterior part of the midgut showed no continuous layer of electron-dense material. Denser accumulations of material were found only at the tips of the microvilli. The contents of vesicles in the epithelium could be labeled with wheat germ agglutinin (WGA)-gold conjugate (Fig. 30b). The labeling could be abolished almost completely by competition with triacetyl chitotriose, but not with N-acetylglucosamine; this shows that chitin or chitin precursors might be present in these vesicles.

The numerous peritrophic membranes which are secreted by the epithelium of the posterior part of the midgut show only a random texture of microfibrils (Peters 1967b). They are used to envelop the remnants of food which are transported from the anterior into the posterior part of the midgut. In this way an unusually large fecal pellet is formed. It has the same size and form as the posterior midgut. The peritrophic envelope is up to 25 mm thick. There is no evidence that in this part of the midgut peritrophic membranes are secreted in a certain time interval according to the digestive cycle as in the anterior part of the midgut. The fecal pellet is dehydrated and finally voided via the very short hindgut (Fig. 30a).

In insects similar contributions of the midgut caeca to the production of peritrophic membranes by the midgut epithelium have been reported.

In the orders Saltatoria and Blattaria conspicuous extensive caeca occur which are evaginations of the anterior midgut. In *Locusta migratoria migratorioides*, *Schistocerca gregaria* (Baines 1978; Bernays 1981) and *Periplaneta americana* (Lee 1968), all caecal cells are involved in the formation of peritrophic membranes. These do not differ morphologically from those formed by the midgut epithelium and join the peritrophic envelope.

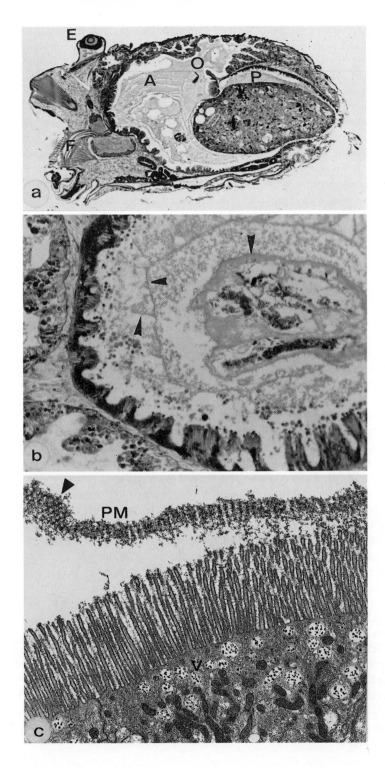

In Acrididae there are six caeca with anterior and posterior protrusions or arms which are lined with typical midgut epithelium. Bernays (1981) found in cross-sections of the posterior arms of *Schistocerca gregaria* that the epithelium on the side adjacent to the midgut is flat and continues without interruption into the distal part of the caecum. On the outer side of the caecum the epithelium forms numerous pockets; these are absent for instance in *Locusta migratoria*, *Melanoplus sanguinipes* and a related species, *Anacridium melanorhodon*. The whole epithelium of these caeca, including the pockets, is able to secrete peritrophic membranes. If the membrane is carefully drawn out of the caecum, those parts which were secreted by the pockets appear as mushroom-like protrusions. In *Locusta migratoria migratoroides* the caeca do not form a complete tube, but strands. These pass into the midgut and join the membranes which are formed by the epithelium of the midgut (Baines 1978). No fundamental, ultrastructural differences could be observed in peritrophic membranes from the caeca or different parts of the midgut. An actively feeding locust secretes a peritrophic membrane at intervals of approximately 15 min. Baines (1978) found that in *Locusta migratoria* the number of peritrophic membranes in the peritrophic envelope depends on the nutritional condition of the locust.

For functional aspects, the elimination of tannin and plant toxins, see Sect. 6.5.1.

Similar conditions were found by Lee in the cockroach *Periplaneta americana* where the midgut epithelium in front of, in and behind the caeca contributes to the production of peritrophic membranes. If the membranes are carefully eased of the anterior part of the midgut of nymphs as well as of adults, those parts which were secreted by the eight caeca appear as eight points like a crown (Fig. 31). As these membranes pass into the midgut they join the peritrophic membranes which were secreted by the midgut epithelium to form a peritrophic envelope. The latter has a thickness between 20–40 µm and consists of at least ten peritrophic membranes. However, these data from light microscopy as well as the measurements of single membranes should be controlled by electron microscopy.

The caeca of mosquito larvae are situated behind the cardia. There are eight dead-end tubes which empty into the anterior end of the midgut, immediately behind the cardia (Figs. 32, 33). In larvae of *Aedes aegypti* and *Culex pipiens fatigans* (*C. quinquefasciatus*), the lateral caeca are the largest, whereas the ventral and dorsal caeca are rather small (Christophers 1960). In *Anopheles stephensi* the caeca are subdivided into a small anterior and a larger posterior part (Imms 1907; Jones 1960).

◄————————————————————————————————

Fig. 29a-c. Formation of peritrophic membranes in harvestmen (Chelicerata, Phalangida). *a* Sagittal section of *Leiobunum rotundum*. The spacious midgut is divided into an anterior (*A*) and a posterior part (*P*) by a dorsal fold of the midgut epithelium. Peritrophic membranes are secreted by the epithelium of both parts. In the posterior part large fecal pellets are formed. *E* Eye; *O* opening of the midgut caeca; *F* foregut; 10x. *b* Transverse section of the anterior midgut where the food residues are surrounded by peritrophic membranes (*arrowheads*); 180x. *c* Electron micrograph showing a newly formed thick peritrophic membrane (*PM*) with a hexagonal texture of microfibrils in the anterior midgut epithelium of *Phalangium opilio*. The formation of peritrophic membranes is most effective between 1 and 3 h after feeding. Note the vesicle *V* in the apical part of the midgut cell; 9000x. *a,b* After Peters (1967b); *c* after Becker and Peters (1985)

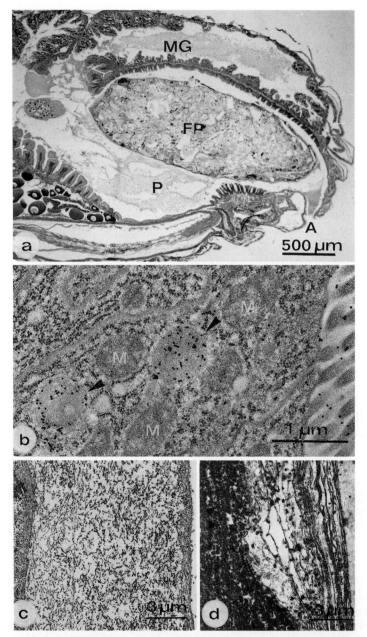

Fig. 30a-d. Formation of fecal pellets in harvestmen, *Phalangium opilio* (Chelicerata, Phalangida). *a* A large fecal pellet (*FP*) is formed in the posterior part (*P*) of the midgut. *A* Anus; *MG* midgut gland. *b* Material for the formation of peritrophic membranes is provided by the secretion cells. N-acetylglucosamine residues which probably belong to chitin precursors can be localized with a lectin, wheat germ agglutinin labeled with colloidal gold as an electron-dense marker (*arrowhead*). *M* Mitochondria. *c* In early stages of fecal pellet formation the peritrophic envelope consists of several light peritrophic membranes with masses of fibrous material in between. *d* In later stages the material condenses to form aggregations of membranes (After Becker and Peters 1985)

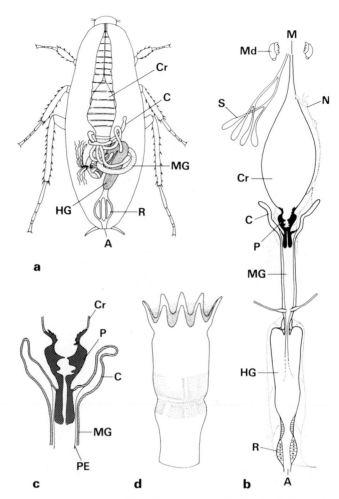

Fig. 31a-d. Periplaneta americana. a Survey of the alimentary canal (after Sanford 1918).
b Alimentary canal (simplified after Bignell 1977). *c* Crop, proventriculus, midgut, and
formation of peritrophic membranes in the caeca (redrawn after Lee 1968). *d* A peritrophic
envelope consisting of many peritrophic membranes is formed in the 8 caeca (after Lee 1968).
A Anus; *C* caecum; *Cr* crop; *HG* hindgut; *M* mouth; *Md* mandible; *MG* midgut; *N* nerve;
P proventriculus (gizzard); *PE* peritrophic envelope; *R* rectal pad; *S* salivary gland

The fine structure of the caecal epithelium was investigated in larvae of *Aedes aegypti*
by Jones and Zeve (1968), Sutter (1977) and Volkmann and Peters (1989a,b). In
contrast to Jones and Zeve, who described only a single cell type, Sutter found two,
and Volkmann and Peters four different cell types. Resorbing/secreting cells
predominate in the epithelium of the caeca and the posterior midgut (Sutter 1977).
They are characterized by the presence of glycogen and lipids, which suggests a
function as storage cells. Some of these cells also contain considerable quantities of
rough endoplasmic reticulum, which may indicate that they synthesize and secrete

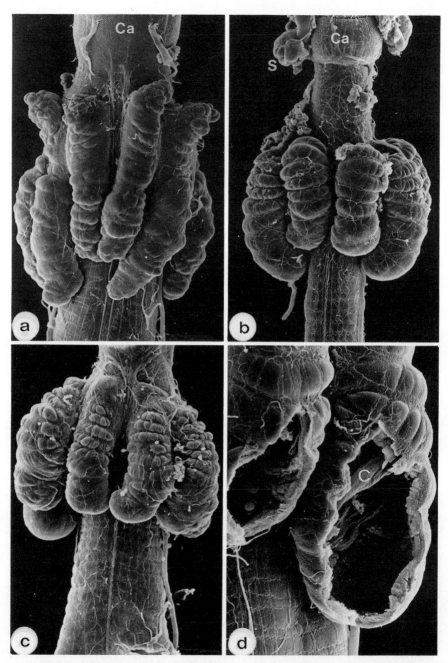

Fig. 32a-d. Scanning electron micrographs of the caeca of larvae of *a Anopheles stephensi*; *Aedes aegypti*; *c Culex pipiens*; *d C. pipiens*: the rear end has been torn off to show the caecal membrane (*C*). *Ca* Cardia; *S* salivary gland (After Volkmann and Peters 1989a)

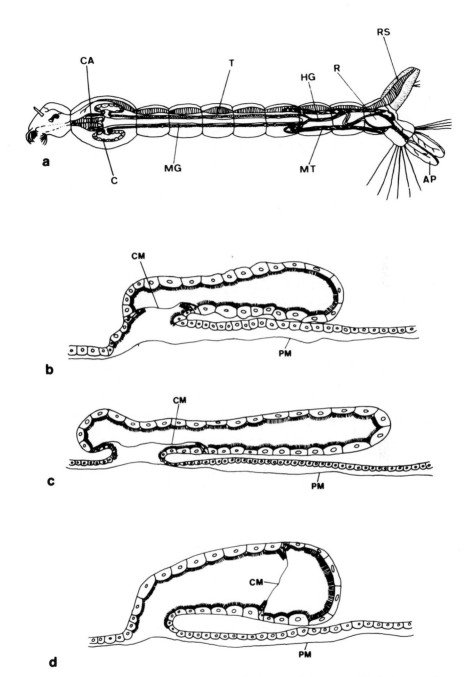

Fig. 33a. Schematic representation of a mosquito larva and the caeca of *b Aedes aegypti, c Anopheles stephensi, d Culex pipiens.* *AP* Anal papillae; *CA* cardia; *CM* caecal membrane; *HG* hindgut; *MG* midgut; *MT* Malpighian tubules; *PM* peritrophic membrane; *RS* respiratory sipho; *T* trachea (After Volkmann and Peters 1989a)

either enzymes or other substances. A second, abundant type of cell in the epithelium of the caeca of mosquito larvae has been called ion-transporting cells as these cells resemble other insect cells involved in osmoregulation. Ion-transporting cells are characterized by the close association of mitochondria and the membrane of the microvilli (Cioffi 1984). A third type of cell is observed occasionally in the basal parts of the epithelium of the caeca. These are very small cells, which occur between resorbing/secreting cells and ion-transporting cells. They are characterized by very electron-dense cytoplasm containing many free ribosomes, occasional mitochondria, and Golgi fields. Vesicles or cytoskeletal structures could not be observed. As Richards (1975) pointed out that there is no replacement or degeneration of cells in the midgut of the caeca of mosquito larvae until metamorphosis, these cells in the caeca were called imaginal cells instead of regeneration cells (Volkmann and Peters 1989a).

Two to three annular rows of a fourth type of cells in the caecal epithelium occur either near the opening of the caecum between midgut and the first resorbing/secreting cells, as in *Ae. aegypti* and *An. stephensi*, or near the posterior end of the caecum among resorbing/secreting cells, as in *C. pipiens* (Fig. 34). These cells protude into the lumen of the caecum and secrete a membrane which has been called the caecal membrane (Volkmann and Peters 1989a). The fine structure of these cells is similar in the three species mentioned above. They have stumpy microvilli, which are arranged irregularly with larger spaces between them, and which are aligned with their tips to the posterior end of the caecum. A finely granular material is secreted by these cells which forms a thin membrane. This is only about 160 nm thick and therefore about five times thinner than the peritrophic membrane of the respective species. It appears to be amorphous and contains cellular debris near the formation cells. The caecal membrane can be labeled specifically with wheat germ agglutinin gold. Competition with 0.3 M N-acetyl glucosamine is ineffective, but competition with 10 mM triacetyl chitotriose reduces the label almost completely (see Sect. 5.2). In order to determine the presence of aminopeptidases, whole midguts or frozen sections of the posterior and anterior midguts, including the caeca, were incubated in a medium containing the substrate L-leucyl-4-methoxy-2-naphtylamide HCL according to Geyer (1973). The cells of the anterior midgut never showed any aminopeptidase activity. A positive reaction was limited to the caecal cells and the posterior midgut cells. The resorbing/secreting cells of the caeca showed a very strong reaction in the apical region and brush border. The ion transporting cells in the caeca of larvae of *Ae. aegypti* and *C. pipiens* reacted less clearly.

The caecal membrane divides the caecum into two compartments. It may function as a barrier to retain cellular debris within the caecal lumen and may also help to maintain the pH gradients between anterior midgut and caecal lumina (Dadd 1975). The caecal membrane does not function as a peritrophic membrane. For functional aspects, see Sect. 6.5.2.

The ability to form peritrophic membranes is restricted in some insects to certain stages: In the caddisfly (Trichoptera), *Limnephilus stigma*, and in the flea (Siphonaptera), *Ceratophyllus sciurorum*, only larvae but not the adults are able to develop a peritrophic envelope (Chaika 1980a,b).

Insects may have a large esophageal invagination, as for instance Odonata like *Aeschna* and *Libellula* (Aubertot 1934), larvae of Lepidoptera (Bordas 1911), and

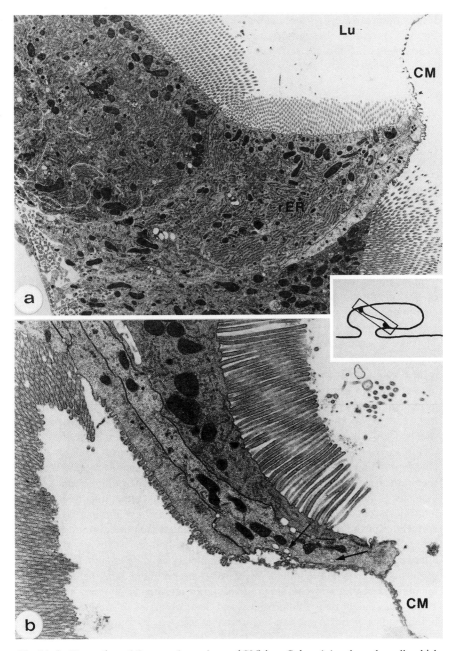

Fig. 34a,b. Formation of the caecal membrane (*CM*) in a *Culex pipiens* larva by cells which protrude into the lumen (*Lu*) of the caecum. These cells are situated among resorption/secretion cells. *Arrows* point to vesicles. *rER* Rough endoplasmic reticulum. *Inset:* Area of section within the caecum. *a* 4600x; *b* 11 000x (After Volkmann and Peters 1989a)

Hymenoptera like *Apis, Bombus* (Wigglesworth 1930) and *Vespa* (Aubertot 1932a,b; Fig. 35). These and other authors assumed that such species may have a double origin of peritrophic membranes, combining the features of Types I and II of Wigglesworth. It was suggested that this may represent a stage in the evolution of Type II from Type I.

In several species of Hemiptera (Heteroptera and Homoptera) a peculiar type of membrane has been described from the midgut lumen (see Chap. 2). They seem to contain no chitin. Marshall and Cheung called these membranes a "plexiform surface coat", whereas others called it "extracellular membrane layers (ECML)" (Fig. 5). Most authors assume that this is a modified form of peritrophic membranes. *Rhodnius prolixus* is the most intensely investigated species. The membranes are formed by the epithelium of the entire midgut of all stages (Lane and Harrison 1979). The anterior midgut functions, according to ultrastructural (Bauer 1981) and physiological studies (Farmer et al. 1981; Barrett 1982), as a storage organ for the ingested blood, as the site of water absorption and urea recycling. The posterior midgut or intestine is divided into two parts differing in diameter, structural features and function, an anterior secretory and a posterior absorptive region. The midgut cells of unfed bugs possess a single plasma membrane with a surface coat. A duplication of this membrane occurs between 12–24 h after a blood meal. By 2 days after engorgement delamination of an outer membrane occurs (Billingsley and Downe 1983). Including the surface coat, these membranes have a diameter of about 10 nm. They are separated by a space of about 10 nm. After lanthanum treatment during fixation the space between the inner and outer plasma membrane of the microvilli is filled with lanthanum which is excluded by columns. These run either diagonally or parallel to the long axis of the microvilli. The columns seem to be associated with intramembranous particles. Each microvillus is surrounded by an irregular spoke-like network which appears to consist at first sight of fibrils, but which are probably sheets. They are connected with a tubular sheath which runs parallel to the double membrane of the microvillus. This sheath may be shared by several microvilli. It is always present, whereas the spokes may be absent. Freeze-fracture replicas revealed that both structures appear to consist of rather large globular units, about 8 nm in diameter (Lane and Harrison 1979). In the less flattened epithelium of the anterior intestine the outer membranes are retained on the microvilli, whereas in the flattened epithelium of the posterior intestine the delaminated membranes tend to be discharged into the lumen where also multimembrane-bound vesicles occur. Ten days after engorgement the outer membranes completely separate the inner membranes of the microvilli from the lumen. Twenty days after a blood meal the delaminated membranes begin to degenerate in the posterior intestine (Billingsley and Downe 1983). Such membrane systems have also been observed in the closely related, blood-feeding South American bug *Triatoma infestans* by Burgos and Guitiérrez (1976), Guitiérrez and Burgos (1978). The flagellate *Blastocrithidia familiaris* is found in the midgut of the latter species. Attachment of these parasites occurs by interdigitation of expanded flagella over and between the microvilli. No attachment to microvilli was observed where the peculiar extracellular membranes are formed by delamination of the outer membrane of microvilli.

Nothing is known about a similar correlation between food uptake and formation of double plasma membranes in sap-feeding Hemiptera and Homoptera. How-

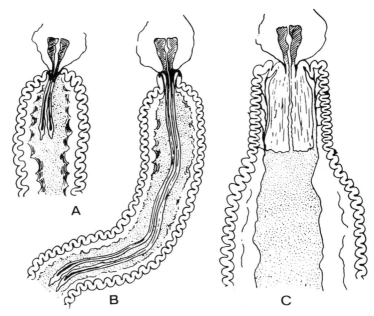

Fig. 35A-C. A comparison of the length of the valvula cardiaca in *A Apis mellifica*; *B Vespa germanica*; and *C Bombus terrestris* shows that *V. germanica* has an extremely long valvula with a length of nearly 6 mm (After Aubertot 1934)

ever, similar structures could be demonstrated in the milkweed bug, *Oncopeltus fasciatus,* by Baerwald and Delcarpio (1983) (Fig. 5), and in *Lygaeus pandurus* (Tieszen et al. 1986).

In *Nepa cinerea* lanthanum applied during fixation did not reach the space between the inner and outer plasma membrane of the microvilli probably because it could not penetrate the delaminated membrane.

Much must still be done for a better understanding of the functional implications of these membranes and their possible relation to peritrophic membranes.

The controversies in the large amount of literature on peritrophic membranes in the honeybee, *Apis mellifica,* had several reasons: (1) It was difficult to obtain a satisfactory fixation; (2) the resolution of the light microscope was inadequate; (3) the chitosan test was a new and controversial method; (4) it was generally believed that cells of endodermal origin were unable to synthesize and secrete chitin; (5) several authors wanted to demonstrate that peritrophic membranes are secreted by a collar of cells around the base of the esophageal valve. Electron microscopy helped to end this confusion and has shown that the peritrophic membranes of the honeybee are secreted by the whole midgut epithelium. However, controversial reports still occur: Cruz Landim (1985) maintained that only specialized cells of the anterior midgut are responsible for the formation of peritrophic membranes, whereas Pabst et al. (1988) obtained histochemical evidence that the whole midgut epithelium is involved in the formation of these membranes. For a review of the older literature, see Aubertot (1934) and Schreiber (1956). The number of peritrophic membranes increases in the honeybee after hatching to seven or eight (von Dehn 1933) or to ten (Weil 1936). However, eventually the number of membranes is by far greater than von Dehn assumed. Weil found more than 20 layers, which is confirmed by the preparation of fresh material. Smaller numbers seem to be a light microscopic fixation artifact. Hering's micrograph (1939) shows that in hatching bees the secretion of peritrophic membranes at the entrance of the midgut is supported already by the secretion of membranes by other midgut cells. Hering distinguished between membranous and secretion-like membranes surrounding food and food residues. The large amount of enzymes which adhere to these peritrophic membranes (Peters and Kalnins 1985) were detected already by Petersen (1912; see Chap. 6.3). Weil (1936) believed that either the whole or at least the peripheral part of the peritrophic envelope was without peritrophic membranes and consisted only of mucous substances. Nevertheless, Campbell (1929), Wigglesworth (1930) and von Dehn (1933, 1936) could show that the peritrophic membranes of the honeybee contain chitin. However, Wigglesworth remarked that he had "not actually demonstrated that the inner layers are chitinous and the outer later layers not chitinous".

Besides, Aubertot (1934) showed that the valvula cardiaca of related species is extremely long: in the wasp, *Vespa germanica*, it has a length of about 6 mm and in the bumblebee, *Bombus lapidarius*, it reaches one-eighth of the length of the midgut (Fig. 35).

Larvae of the honeybee have been investigated by several authors with controversial results (Anglas 1901; Rengel 1903; Evenius 1925; Nelson 1924; Oertel 1930; Aubertot 1932a,b; Kusmenko 1940). According to Kusmenko, two types of peritrophic membranes are secreted in different parts of the midgut: (1) a thick granular layer is secreted by the cells at the entrance of the midgut ("Promesenteronring"). (2) Membranes are secreted as slimy pieces of different thickness by the remaining midgut epithelium. Both types were believed to contain no chitin, but no test was made to control this assumption. Light and electron microscopic investigations by Schmitz (unpubl. results) revealed that the formation zone in a fold at the entrance of the midgut consists in young larvae of about 15 rings of cells (Fig. 36b,c) which together have a width of about 100–150 µm. These cells secrete a peritrophic envelope with a thickness of about 5 µm. It contains several peritrophic membranes and forms a sac which is filled with food and food remains; mainly pollen could be distinguished. With increasing age the formation zone is enlarged by an upfolding of the epithelium. The cells are smaller than in young larvae and have extremely long microvilli. Their length is about 150 µm, the diameter at the base is about 180 nm and in the distal parts about 26 nm. Between these microvilli sections of microfibrils can be found in electron micrographs. In the light microscope the long microvilli and the secreted material appear to stream out of the formation cells like a waterfall (Fig. 36b,c); see also the micrographs given by Aubertot (1934) and Kusmenko (1940). In older larvae the sac-like peritrophic envelope appears to consist of several layers of different appearance. Schmitz distinguished up to five layers, each containing a different number of peritrophic membranes. However, these layers may be due to different textures of microfibrils and a different amount of adsorbed enzymes. Kümmel (1956) found no microfibrils. Peters (1969) showed a random texture, and Schmitz (unpubl. results) reported a hexagonal texture of microfibrils with a diameter of the honeycomb pattern of 220 nm. It seems that the random texture is reinforced as usual by a hexagonal network of microfibrils.

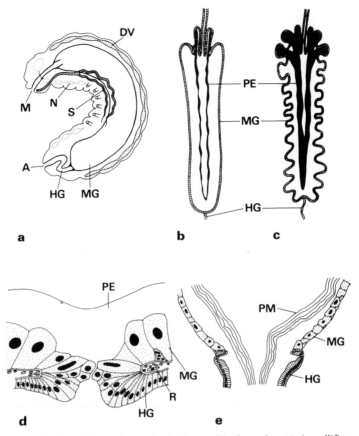

Fig. 36a-e. Formation of peritrophic membranes by the larva of the honeybee, *Apis mellifica*. *a* Survey of the alimentary canal. *A* Anus; *DV* dorsal vessel; *HG* hindgut; *M* mouth; *MG* midgut closed against hindgut; *N* ventral nerve cord; *S* salivary gland (after Schoenichen 1930). *b,c* Schematic representation of the closed midgut and sac-like peritrophic envelope in *b* a young and *c* an old larva. *HG* Hindgut; *MG* midgut; *PE* peritrophic envelope (after Schmitz, unpubl. results). *d,e* Rupture of the midgut-hindgut closure at the end of the feeding stage in the wasp, *Vespa germanica* (redrawn after Rengel 1903). *d* Section through the border between midgut (*MG*) and hindgut (*HG*) epithelium. *PE* Peritrophic envelope consisting of many peritrophic membranes. *R* Regeneration cell. *e* After the feeding stage has ended, the closure between midgut and hindgut is ruptured, the peritrophic envelope is moved into the hindgut and voided together with the feces. *PM* Peritrophic membranes

A peculiar variant is represented by those species in the larvae of which there is no connection between the midgut and the hindgut; in other words, where the midgut is like a sac. This has been found in the larvae of almost all species of the Hymenoptera Apocrita (Rengel 1903; Aubertot 1932a,b; Kojima 1983).

During the larval stages a single layer of cells closes the hind part of the midgut (Rengel 1903) (Fig. 36d) thus preventing the transfer of digested material from the midgut to the hindgut. In *Vespa germanica* the hind part of the midgut and the hindgut are connected by a short and slightly bended stalk of cells (Rengel 1903). As there are regeneration cells in the stalk, Rengel assumed that this region belongs to the midgut which is only narrowed and without a lumen in this part. Midgut and hindgut are separated only physiologically. At the end of the feeding period the wasp larva closes its honeycomb cell and tries to get rid of the contents of the midgut. The peritrophic envelope with the food residues is pressed against the closed hindparts of the midgut by violent muscle contractions. This causes the formation of a lumen in the stalk and the passage of the midgut contents into the hindgut (Fig. 36e). Then the feces are voided into the honeycomb cell. This takes several hours. The sticky material is pressed against the bottom of the cell to form a little bowl. In autumn the number of bowls in a given cell indicates the number of generations reared in this cell. Frenzel (1886) observed in larvae of the honeybee, *Apis mellifica*, that about 30–40 pellets are voided by the larva. In this species the fecal pellets seem to be removed by worker bees.

Among the social wasps, the 5th instar larvae of the Vespinae and most Polistinae eject the meconium covered with peritrophic membranes just before pupation and compress it to the bottom of their cells (Jeanne 1980; Kojima 1983). In the Polistinae females gnaw small holes with a diameter of about 1 mm through the bottom of the cell, 1–3 days after the larva has spun a cocoon. They extract the peritrophic sac through the hole and deposit it in the vicinity of the nest. Afterwards they close the hole with saliva which forms a semitransparent membrane. Holes of old cells are larger than those of young cells because the cells are used several times (Kojima 1983).

Among adult Diptera, mosquitoes (Culicidae) of medical importance were investigated very often, blackflies (Simuliidae) and sandflies (Phlebotomidae) less frequently with respect to the formation of peritrophic membranes.

In adult mosquitoes (Culicidae) peritrophic membranes are secreted only by females after a complete blood meal. They are lacking in females after the ingestion of nectar, after an incomplete blood meal or after the ingestion of blood and nectar. They are also lacking in males which feed only on nectar. Stohler (1957) and Freyvogel and Stäubli (1965) suggested that the whole midgut epithelium of females is able to secrete peritrophic membranes, but that the quantity is less in the anterior part than in the posterior part of the midgut. In *Aedes aegypti* the secreted material from the anterior part forms a rostral whorl, and in *Anopheles gambiae* secretion in this part lasts only for a short time (Stohler 1957). In *Aedes triseriatus* about 20 min after feeding membranous material has been secreted in the anterior part of the midgut; it is connected with similar material in the posterior midgut (Richardson and Romoser 1972). Both parts gave a positive chitosan reaction and therefore should be peritrophic membranes. Fifty minutes after feeding the blood bolus is enclosed in a developing peritrophic envelope consisting of numerous peritrophic membranes. Then its anterior portion is transported posteriorly by the peristaltic movements of the gut, until by 4 h after feeding it forms a wrinkled mass in the region of the junction between the anterior and posterior part of the midgut. The wrinkled mass corresponds to the whorl described by Stohler. Richardson and Romoser suggested that it might act as a plug which could separate both parts of the midgut and allow it to take a subsequent sugar meal into the anterior part without interference with the digestion of the blood meal in the posterior part of the midgut.

Waterhouse (1953a) and Zhuzhikov (1962) assumed that peritrophic membranes are formed only in the posterior midgut of mosquitoes. After a blood meal the ingested blood is pumped immediately into this part of the midgut. This is in accordance with a more recent electron microscopic investigation of the formation of peritrophic membranes in *Anopheles stephensi* by Berner et al. (1983) (Figs. 37a, 38a). Twelve hours after a blood meal peritrophic membranes appeared, and 48 h after feeding the peritrophic envelope was completed. It consisted of several peritrophic membranes which were closely apposed to the blood mass. During digestion the volume of the contents in the posterior midgut, the so-called stomach, decreased considerably. The excess material of peritrophic membranes formed a plug between the anterior and posterior part of the midgut (see above) as well as between the latter and the hindgut. It could separate these parts functionally as Berner et al. pointed out. At the end of digestion, 60–72 h after feeding, the remains of the blood meal were voided together with the peritrophic membranes.

Recently, Perrone and Spielman (1986) interpreted these features as follows. The most anterior and posterior parts of the midgut are less distensible than the median part. Therefore, in these parts a columnar epithelium persists, whereas after engorgement with blood the median part of the midgut epithelium is flattened to a squamous epithelium. The columnar epithelium is able to secrete a larger plug-like mass of peritrophic membranes which was described as whorls by Stohler (1957). Moreover, Perrone and Spielman could show by autoradiography that the assumptions of Waterhouse (1953a), Zhuzhikov (1962) and Richards and Romoser (1972) concerning the formation of peritrophic membranes in restricted parts of the midgut are wrong. They injected tritiated glucosamine into female mosquitoes immediately after a blood meal. The autoradiograms proved that the whole midgut epithelium is involved in the secretion of peritrophic membranes.

The description of the appearance of a distinct peritrophic envelope surrounding a blood meal is controversial in light microscopic investigations. Waterhouse (1953a) found in *Aedes aegypti* and *Culex fatigans*, and Stohler (1957) in *Aedes aegypti* a gelatinous zone shortly after feeding. A thin but tough transparent membrane which enclosed the blood bolus could be isolated 24 h after feeding. Zhuzhikov (1962) found a distinct peritrophic membrane in *Aedes aegypti* 20 min after feeding, which was confirmed for the same species by Kusnetsova (1968), and for *Aedes triseriatus* by Richardson and Romoser (1972).

A first electron microscopic study by Bertram and Bird (1961) revealed that in *Aedes aegypti* 30 min after a blood meal fine, particulate material appears between the microvilli and beyond their tips. At about 2 h an extensive layer of amorphous material had developed. Secretion of peritrophic membranes continued until 24 h after feeding and resulted in a multilayered peritrophic envelope with a thickness of 0.4–1.2 µm.

Perrone and Spielman (1988) tried to solve the problem more accurately. They fed ferritin together with blood to female *Aedes aegypti*. Two hours after engorgement the ingested erythrocytes were found with a distorted surface and tightly appressed in the periphery of the blood bolus. The space between the rim of the blood bolus and the microvilli was filled with an amorphous mass which contained ingested ferritin. Moreover, the ferritin appeared also between the microvilli. This means that during the first 2 h after engorgement no material could have been secreted which could give rise to peritrophic membranes. In another experiment these authors injected tritiated glucosamine into mosquitoes immediately after a

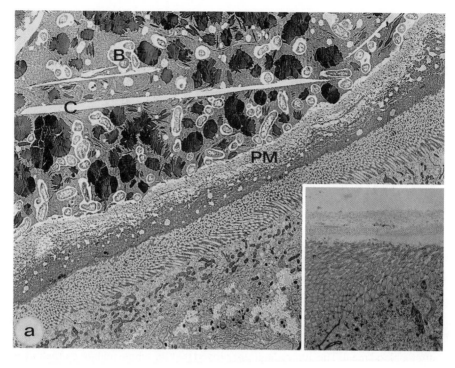

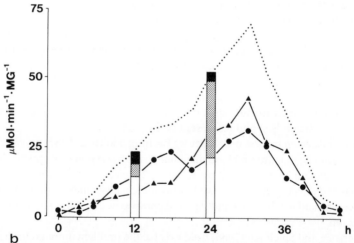

Fig. 37a. Section through the midgut of a female *Anopheles stephensi* (Diptera, Culicidae) 48 h after a blood meal. The formation of a peritrophic envelope made up of many peritrophic membranes *(PM)* is completed. The blood clot contains many bacteria *(B)* and hemoglobin-like crystals *(C)*;4000x. *Inset:* Peritrophic envelope formed 20 h after an enema of physiological salt solution; 8600x (after Berner et al. 1983). *b* Effect of a blood meal on aminopeptidase activity in different parts of the midgut of female *Aedes aegypti.* Five midguts were separated into epithelium (•) and contents (▲). Additionally, five midguts were dissected at 12 and 24 h after a blood meal: *open columns* epithelium; *stippled columns* digestive juice; *dark columns* blood bolus (After Graf and Briegel 1982)

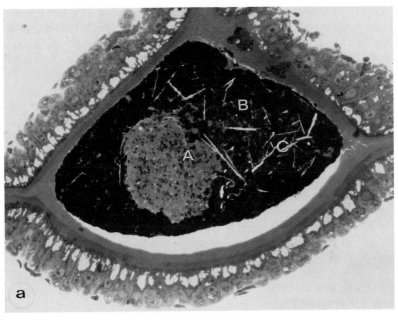

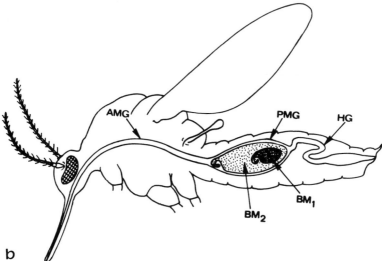

Fig. 38a. Section through the midgut of a female *Anopheles stephensi* (Diptera, Culicidae) 30 h after a blood meal (*B*) which had been taken 18 h after an enema of physiological salt solution. Remnants of the peritrophic envelope (*A*), formed after the enema, are surrounded by the blood clot with hemoglobin-like crystals (*C*) of the ensuing blood meal (*B*). No protease activity was found 10–30 h after an enema. Nevertheless, a peritrophic envelope was formed; 320x (after Berner et al. 1983). *b* If the remnants of a former blood meal (*BM1*) have not been voided before a new blood meal occurs, these are surrounded by the new blood clot (*BM2*) and wrapped in a new peritrophic envelope. *AMG* Anterior midgut; *PMG* posterior midgut; *HG* hindgut (After Waterhouse 1953a)

blood meal. During 2 h after application, the radioactive material appeared in the midgut epithelium but was still confined to the microvilli. Between 4–8 h after application radioactive material was secreted into the space between microvilli and the rim of the blood bolus. By 12 h the labeled material formed a discrete and uniform layer surrounding the blood bolus completely. Electron microscopically a peritrophic envelope consisting of many peritrophic membranes could be discerned 8 h after a blood meal had been consumed. The first appearance of a distinct peritrophic envelope seems to be specific for the different species of mosquitoes. It is of great importance for the suitability of a given species of mosquitoes as a vector for a *Plasmodium* species (see Sect. 6.6). Freyvogel and Stäubli (1965) observed a distinct peritrophic envelope in *Aedes aegypti* 5–8 h, in *Anopheles gambiae* 13 h, and in *Anopheles stephensi* 32 h after feeding. *Anopheles maculipennis atroparvus* did not form a peritrophic membrane. During the digestion of a blood meal the peritrophic membranes become more and more fragile (Stohler 1957). When a second blood meal is taken before the first is digested, the second meal partly or completely surrounds the first, and a new peritrophic envelope is secreted around it (Waterhouse 1953a; Fig. 38b).

Culex species seem to have the same time frame of peritrophic membrane formation, as could be revealed by light and electron microscopy (Whitfield et al. 1973; Houk et al. 1979). Secretion of precursor material for peritrophic membranes with a thickness of about 100 μm occurred 8–12 h after the ingestion of blood, and a multilayered and fibrous peritrophic envelope was observed 20–24 h after a blood meal. Between 48–72 h after the blood meal the midgut had been cleaned from all remnants of blood and peritrophic membranes. In *Culex nigripalpus* Romoser and Cody (1975) observed that blood occurs in the anterior part of the midgut for about 4–5 h after a blood meal. Subsequently, blood and newly formed peritrophic membranes are moved posteriorly and the membranes form a plug which closes the anterior part of the posterior midgut. As it is possible during dissection of mosquitoes to pass blood into the hindgut, the peritrophic envelope must be perforated at its posterior end. In this species the peritrophic envelope is retained in the posterior midgut even after the remnants of the digested blood have been voided through the hindgut between 72–187 h after engorgement.

In adult blackflies peritrophic membranes are formed only by females after the ingestion of blood (Reid and Lehane 1984), but not after feeding a mixture of 70% blood in 1 mol/l sucrose solution (Yang and Davies 1977). Before blood feeding no peritrophic membranes could be observed in the empty lumen of the midgut. Immediately after engorgement the secretion of peritrophic membranes started in the entire midgut epithelium until about 2 h later. Twenty-four hours after a blood meal the peritrophic envelope seemed to be fully differentiated. The number of distinct layers appears to be species-specific. The peritrophic envelope of *Simulium damnosum* has 15 layers which are formed within 90 min after a blood meal (Lewis 1953), that of *S. aureum* has 7 layers (Yang and Davies 1977), that of *Odagmia ornata* has 5 layers, that of *S. lineatum* has 4 layers. The innermost layer, next to the microvillous border, stained conspicuously electron dense. In *S. equinum* only a single membrane with interruptions was observed (Reid and Lehane 1984). *S. venustum*, a mammolophilic species, formed a thinner peritrophic envelope than six ornithophilic species (*S. aureum, S. coxtoni, S. quebecense, S. rugglesi, Prosimulium decemarticulatum*) (Yang and Davies 1977). In other species peritrophic envelopes of varying thickness have been observed (Fallis 1964). Digestion of blood needed 120–180 h. After that time the peritrophic envelopes disintegrated gradually, starting at the posterior end.

In adult sandflies, *Phlebotomus longipes*, a peritrophic envelope surrounds the blood meal in the posterior part of the midgut 24 h after ingestion (Gemetchu 1974). It becomes thicker and stronger by the secretion of further peritrophic membranes until 48 h after feeding. Then its thickness is of the order of 1–5 μm. It is confined, except for a short anterior extension, to the posterior part of the midgut, and it seems to be tight at both ends. The peritrophic envelope did not dissolve in distilled water, saline or KOH. The chitosan test was negative. But microfibrils were observed and therefore chitin should be present. About 3 days after feeding breakdown of the peritrophic membranes starts at the anterior end of the envelope and proceeds on the following 2 days. Digestion of blood ends at about the sixth or seventh day after feeding. Then the residues of the blood meal and the peritrophic membranes are voided. In only partially fed sandflies the first blood meal is surrounded by the second meal as described already for mosquitoes.

3.3.3 Secretion of Peritrophic Membranes in Restricted Parts of the Midgut

This has been observed in several groups of animals, mostly in insects. Copepoda and mites have been mentioned already (Chap. 2, p. 13). It has been found that in both groups peritrophic membranes are secreted only in the posterior part of the midgut. In the praying mantis, *Mantis religiosa*, peritrophic membranes are secreted only by the posterior midgut (Kenchington, pers. comm.). Ascidia (Tunicata) are filter feeders which wrap the filtered material into mucus and transport it into the stomach where delicate peritrophic membranes are secreted. These surround the strand of food material.

The ant lion, the larva of *Myrmeleon europaeus*, was investigated by Lozinsky (1921) and Schmitz (unpubl. results). These larvae catch their prey with their long mandibles and inject regurgitated enzyme-containing contents of their midgut through these injection needles into the prey. After extraintestinal digestion has taken place, the liquid material is sucked into the crop of the ant lion. From there it passes into the midgut which consists of two spacious parts connected by a short tube. There is no passage between midgut and hindgut. Secretion of peritrophic membranes occurs in the tube-like median part and in the large posterior part of the midgut. The membranes show a hexagonal texture of microfibrils, the dimensions of which may be correlated with those of the microvilli. Peritrophic membranes are secreted independently of the ingestion of food. Specimens which had starved for 50 days, nevertheless secreted peritrophic membranes in the meantime. Their previous meal was still in the middle part of the midgut where it was surrounded by a thin peritrophic membrane. Usually this part is used only shortly for the passage of food from the anterior to the posterior part of the midgut. Successively secreted peritrophic membranes surround the incoming food as well as the digested remains of the previous meals. The resulting conglomerate can only pass into the hindgut, which occurs probably at the end of pupation. The fecal pellets are voided by the adult after hatching.

Last, but not least, the larvae of some species of Coleoptera should be mentioned. They use peritrophic membranes as coccon material during pupation. This will be described in Section 6.8.

3.3.3.1 Secretion of Peritrophic Membranes by Annular Groups of Cells in the Posterior Midgut

The feeding larvae of two Ptinidae (Coleoptera), *Ptinus tectus* and *Gibbium psylloides*, produce normal peritrophic membranes which are not secreted throughout the length of the whole midgut but only by cells in the middle third of the midgut (Rudall and Kenchington 1973). These membranes contain microfibrils which exhibit all types of textures. They surround the food residues as usual. Towards the posterior end of the midgut fecal pellets are formed. They are attached to a delicate nonchitinous thread which issues from the anus. These threads are never used for cocoon formation. After the last instar larva has ceased to feed, its peritrophic membranes are transformed into cocoon threads (Sect. 6.8).

3.3.3.2 Secretion of Peritrophic Membranes in the Cardia

This is the most successful restriction of the ability to secrete peritrophic membranes. It was called type II by Wigglesworth (1930) and type I by Waterhouse (1953a). Secretion of peritrophic membranes is confined to a small group of cells at the entrance of the midgut. The formation zone lies in a large fold consisting partly of foregut and partly of midgut epithelium. The foregut cells are characterized by a cuticle of different thickness. Wigglesworth (1930) suggested that a viscous material is secreted by the anterior midgut cells which is then molded into a tube-like, single peritrophic membrane by an annular thickening of the cuticle of the valvula cardiaca. For a long time this molding was taken as a characteristic of type II formation of peritrophic membranes. However, in several species it was observed that more than one membrane is formed. Especially electron microscopic investigations could demonstrate that the peritrophic membranes seem to have their definite size and form after passing only a few cells in the formation zone. Therefore, there is no need for a molding mechanism at the exit of the formation zone.

A well-known example of this type of formation is the peritrophic membrane of earwigs. The common earwig, *Forficula auricularia*, was investigated several times already (Schneider 1887; Adlerz 1890; Cuénot 1895, Wigglesworth 1930; Aubertot 1934; Peters et al. 1979). Another species, *Anisolabis littorea*, was investigated by Giles (1965). Until Aubertot (1934) it was assumed that there is only a single type of peritrophic membrane in the common earwig. Since then it could be shown that there are two large annular folds of the anterior midgut epithelium (Fig. 15) in which different types of peritrophic membranes are formed. A few rows of cells of the anterior part of the first fold are able to secrete peritrophic membranes with an orthogonal texture of microfibrils. In negatively stained preparations these microfibrils proved to be bundles of about ten microfibrils Peters et al. 1979). The formation cells usually have long microvilli with a length of about 5 μm. These microvilli are arranged in rank and file; only occasionally a staggered pattern has been observed (Figs. 17, 18). The chitin-containing microfibrils of the peritrophic membranes polymerize between the bases of the microvilli. There is an exact correlation between the pattern of microvilli and the texture of microfibrils. Up to four separate peritrophic membranes have been observed in the microvillous border (Fig. 17). The mechanism of extrusion is unknown. As there are vesicles with different types of contents in the cells of the formation zone, extrusion of the peritrophic membranes could be achieved by the pressure everted by subsequently secreted material. There seems to be no common or sticky matrix material around and between the strands of microfibrils. Usually several peritrophic membranes stick together to form thick and compact aggregated peritrophic membranes (Figs. 18, 19). Together they form the inner peritrophic envelope which surrounds the food residues. It seems to be unlikely that these membranes are tube-like. Probably patches are formed which aggregate to result in a continuous tube-like peritrophic envelope. In a second annular fold of the anterior midgut epithelium (Fig. 15) numerous, very thin peritrophic membranes are secreted. In contrast to the membranes formed in the first fold, these membranes show a random orientation of microfibrils (Fig. 16). Together these peritrophic membranes form the outer peritrophic envelope which has a thickness between 4.5–12 μm. In unfixed material it appears, in contrast to the inner peritrophic envelope, to be loose and sticky. Matrix material seems to be present and therefore the outer, but not the inner peritrophic envelope of the earwig might be a permeability barrier.

Diptera are remarkably heterogeneous in terms of formation and structure of peritrophic membranes. This appears to be a good example of mosaic evolution. It might be the result of highly different specialization in connection with the use of different food resources by larvae and adults. In adult Nematocera and in the Brachycera, as far as they belong to the Tabanomorpha and Asilomorpha, these membranes are formed according to type I. In the larvae of Nematocera and adults of Brachycera Muscomorpha the ability to secrete peritrophic membranes is confined to annular groups of cells at the beginning of the midgut epithelium in the cardia. Nothing is added by the remaining or even the adjacent midgut epithelium. In spite of the variability in membrane structure which is observed in the larvae of Nematocera and the variability of food, the cardia of larvae of Diptera is rather uniform (Aubertot 1934). The term proventriculus should not be confused with the term cardia. A true proventriculus is part of the foregut and provided with cuticular structures which enable the proventriculus to crush food. The cardia of Diptera is a complex structure, consisting of the posterior foregut or valvula cardiaca and the most anterior part of the midgut epithelium (Figs. 39–41, 43, 45, 47, 48). The valvula cardiaca has a cuticle of different thickness, and muscles and compartments filled with hemolymph. Therefore, it is able even to twist along its longitudinal axis during rhythmic contractions as has been observed in Tipulidae by Aubertot (1934). The cuticle of the foregut is of different thickness. It ends as a very thin ramification between the last foregut cell and the first midgut cell. In many species annular thickenings are present in the cuticle of the valvula cardiaca. Wigglesworth assumed that these are used to mold the viscous secretions of the formation cells into a membrane of uniform thickness. However, as mentioned above, electron microscopy has shown that a single membrane or even several membranes of uniform thickness are present already after the first 10–20 cells have been passed by the newly secreted membrane. Therefore, Richards and Richards (1969) suggested that these thickenings may act as a valve in order to prevent food and microorganisms from entering the formation zone, "but conceivably they might have some other functions as pushing the completed peritrophic membrane posteriorly". The entire sequence of membrane secretion can be followed in a single longitudinal section. During formation, the membrane lies on top of the microvilli (Fig. 44). An exception are only the larvae of Tipulidae (Fig. 39) where the membranes could be found in the microvillous border. The first cell in the formation zone has no cuticle and usually long microvilli. Therefore, the junction between foregut and midgut is readily recognizable. Moreover, the internal organization of the cells changes considerably. Especially the first rows of cells in the formation zone are rather sensitive to osmotic stress, like cells in the rectal pads of insects. Therefore, there is always the danger of artifacts, e.g. swollen or detached tips of microvilli. No mitoses have been described in the formation cells. Their number seems to remain constant despite the enormous secretory efficiency. This reminds one of a similar situation in the odontoblasts which secrete continuously the radula of snails (Gastropoda) and other Mollusca. In the cardia the newly formed peritrophic membrane appears as a straight tube. But in the anterior part of the midgut a larger convoluted amount of peritrophic membranes accumulates, e.g. in flies and their larvae, which may serve as a reserve. It can be used if a sudden uptake of larger amounts of food occurs. The Brachycera Muscomorpha are distinguished by a tendency to increase the number of peritrophic membranes from one to three. This is achieved by folding the epithelium of the formation zone. These membranes are characterized by morphological as well as biochemical asymmetries concerning the location of glycoproteins, lectins and proteoglycans (acid mucopolysaccharides) (see Sect. 5.2).

In the majority of the larvae of Nematocera a single peritrophic membrane is secreted. It seems to be remarkable that in species of the most primitive families the formation of several membranes has been observed. In Tipulidae of the genera *Dictenidia, Tipula* and *Ctenophora* the formation of membrane patches in the cardia was described by Peters et al. (1978; Fig. 39). They contain microfibrils arranged in an orthogonal texture. Five such membranes join to form a continuous, tube-like peritrophic envelope in the midgut. Larvae of *Phalacrocera* sp. (Cylindrotomidae) secrete six and Mycetophilidae form several peritrophic membranes in the folded formation zone of the cardia (Aubertot 1934). Larvae of a *Liponeura* species are said to secrete peritrophic membranes in the cardia as well as in the whole midgut epithelium (Aubertot 1934). In a highly specilialized species, *Miastor metraloas* (Cecidomyidae), patches of membrane are secreted which fuse to form a single peritrophic membrane (Peters et al. 1978; Fig. 43b,c).

The peritrophic membranes of larvae of Nematocera are distinguished by a remarkable diversity of structural patterns. These may differ between families or even closely related genera as has been shown in Culicidae. There are either continuous layers of different thickness, or interrupted layers or rod-like structures (Figs. 39–46). The dimensions are compiled in Table 5.

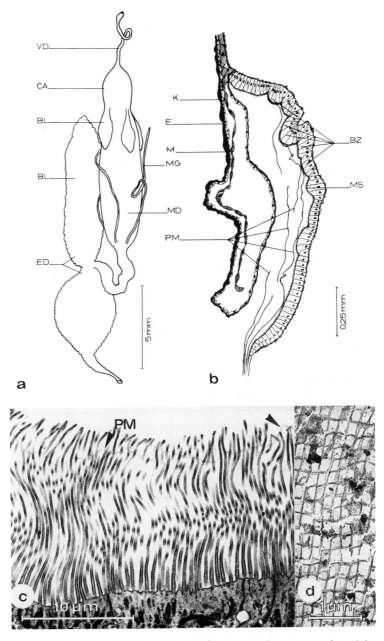

Fig. 39a-d. Larva of *Dictenidia* sp. (Diptera, Tipulidae). *a* Schematic representation of the gut. *b* Schematic representation of the right half of the cardia. *c* A peritrophic membrane leaves the brush border; 3600x. *d* Orthogonal texture of microfibrils as revealed by negative staining; 17 000x. *BL* Caecum; *BZ* formation zone; *CA* cardia; *E* epithelium of the valvula cardiaca; *ED* hindgut; *K* cuticle; *M* muscle; *MD* midgut; *MG* Malpighian tubules; *MS* microvilli; *PM* peritrophic membranes; *VD* foregut (After Peters et al. 1978)

Table 5. Dimensions of peritrophic membranes of larvae of Nematocera

Species	Width (nm)	Width of electron-dense layer (nm)
Psychoda alternata	500–600	
Anisopus cinctus	700	
Bibio hortulanus	1900	
Penthetria holosericea	3500	
Sciara sp.	800	
Aedes aegypti	600–800	200
Anopheles stephensi	400–600	80–100
Culex pipiens	1200–2000	50–80
Chironomus strenzkei	440	110
Chironomus thummi	1700–1800	80
Odagmia ornata	1000–2000	60

Until now only a few species have been investigated using electron microscopy:

Tipulomorpha
Tipulidae
Dictenidia sp.
Psychodomorpha
Psychodidae
Psychoda alternata
Bibiomorpha
Anisopodidae (Phrynidae)
Anisopus cinctus
Bibionidae
Bibio hortulanus
Penthetria (Amasia) holosericea
Sciaridae
Sciara sp.
Cecidomyidae
Miastor metraloas
Culicomorpha
Culicidae
Aedes aegypti
Anopheles stephensi
Anopheles albimanus
Culex pipiens fatigans
Culex restuans
Chironomidae
Chironomus strenzkei
Chironomus thummi
Chironomus thummi piger
Simuliidae
Odagmia (Simulium) ornata
Simulium venustum

In *Psychoda alternata* the fully formed peritrophic membrane has a thickness of 0.5–0.6 µm. On the lumen side it has a thin electron-dense layer which is 20 nm thick. It is reinforced by an electron-lucent, 200-nm-thick layer which contains microfibrils in a random texture. Another electron-dense layer with a thickness of up to 400 nm follows on the epithelium side of the membrane (Fig. 40).

In *Anisopus cinctus* the three larval instars show a different production of peritrophic membranes. The first instar secretes very thin membranes. In the second instar a translucent, elastic tube can be isolated. In the third instar the production of membranes ceases and the hindgut contains only remains of these. Already the first cell in the formation zone secretes material which afterwards becomes a 10-nm-thick, electron-dense layer on the lumen side of the membrane. It is reinforced by less electron-dense material secreted by the fifth cell and following cells. It contains an orthogonal pattern of microfibrils with spacings of 120–130 nm. Finally, a conspicuous electron-dense grid of microfibrils is deposited on the epithelium side of the membrane, continuing the orthogonal texture of the preceding part. It seems that the microfibrils are covered with electron-dense proteoglycans. The fully formed membrane has a thickness of about 0.7 µm.

The peritrophic membrane of *Bibio hortulanus* is characterized by a 100-nm-thick, electron-dense layer on the lumen side. It is followed by an orthogonal pattern of electron-dense rods which have a height of about 200 nm and a width of 60–90 nm (Fig. 42). These are connected with the orthogonal texture of microfibrils which are embedded in a less electron-dense matrix. The spacings are in the range of 105–125 nm as measured from the middle of a bundle to the next. The diameter of the microvilli of the formation cells is in the range of 90–100 nm. After staining with Congo red an orthogonal pattern can be observed in polarized light. Although the texture of microfibrils is below the resolution of the light microscope, the regular pattern seen in polarized light may be correlated with it. It may be transposed to the light microscopic level due to interference. Another regular pattern was also observed in polarized light. Anisotropic crossbands were found which have a width of 60 µm if crossband and intermediate band are added. In this case no structural correlate has been found until now. The fully formed membrane has a thickness of about 1.9 µm. *Penthetria holosericea*, another species of the Bibionidae, has a peritrophic membrane with up to 15 layers of different electron density and a total thickness of about 3.5 µm (Fig. 43). A crossband pattern has been observed as in *Bibio hortulanus*. It has a smaller width of about 40 µm.

In *Sciara* sp. the first four rows of cells in the formation zone secrete a layer of about 500 nm thickness which contains an orthogonal network with spacings of about 10 nm. The following rows of cells add an electron-dense layer with a thickness of about 300 nm. A very fine, crossband pattern was observed with spacings of 10 nm.

In the preceding species only an orthogonal but no hexagonal texture of microfibrils was observed. The peritrophic membranes of Tipulidae were seen in the microvillous border (Fig. 39), but in the remaining species they appeared only on top of the microvilli (Figs. 40–46). Nevertheless, there is a conspicuous correspondence between the diameter of the microvilli and the spacings of the bundles of microfibrils, which is in agreement with the hypothesis of Mercer and Day (1952; see Sect. 3.2):

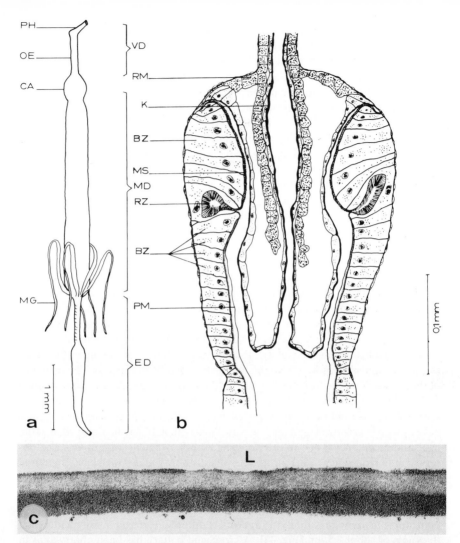

Fig. 40a-c. Larva of *Psychoda alternata* (Diptera, Psychodidae). Schematic representation of *a* the gut and *b* the cardia. *c* On the lumen side (*L*) of the peritrophic membrane there is a thin and on the epithelium side a very thick electron-dense layer. *BZ* Formation zone; *CA* cardia; *ED* hindgut; *K* cuticle; *MD* midgut; *MG* Malpighian tubules; *MS* microvilli; *OE* esophagus; *PH* pharynx; *PM* peritrophic membrane; *RM* circular muscles; *RZ* giant cell with a deep indentation; *VD* foregut; 5000x (After Peters et al. 1978)

Species	Diameter of microvilli (nm)	Spacings of microfibrils (nm)
Dictenidia sp.	160–200	210–240
Anisopus cinctus,	85–90	120–130
Bibio hortulanus,	90–100	105–125
Penthetria holosericea	90–110	120–130
Sciara sp.	80–90	100

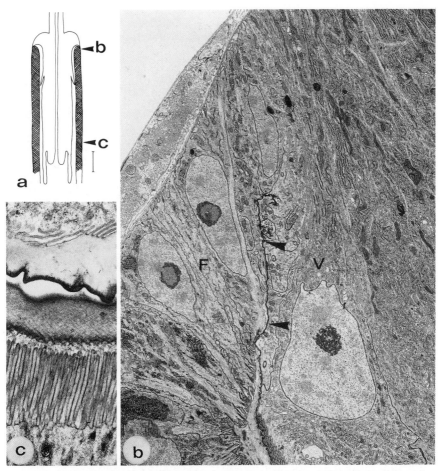

Fig. 41a-c. Larva of *Anisopus cinctus* (Diptera, Anisopodidae). *a* Schematic representation of the cardia indicating the sites of section *b* and *c*. *b* At the end of the valvula cardiaca its cuticle is very thin and marks the border between valvula cardiaca (*V*) and formation zone (*F*) of the peritrophic membrane; 4000x. *c* Fully developed peritrophic membrane with an electron-dense layer on the lumen side and a grid texture which is conspicuously electron dense on the epithelium side; 8000x (After Peters et al. 1978)

In larvae of Culicidae different structural patterns have been described in three genera until now.

In the elongated cardia of the larvae of the yellow fever mosquito, *Aedes aegypti*, a single tube-like peritrophic membrane is secreted by several annular rows of cells belonging to the most anterior part of the midgut epithelium (Fig. 44). The first of these cells is easily recognized as it lacks the cuticle of the adjacent cells of the valvula cardiaca which belong to the foregut. It has microvilli of different length which resemble a paintbrush, with long microvilli in the center and shorter ones at the rims. The second cell may have areas with only short microvilli, but the following cells

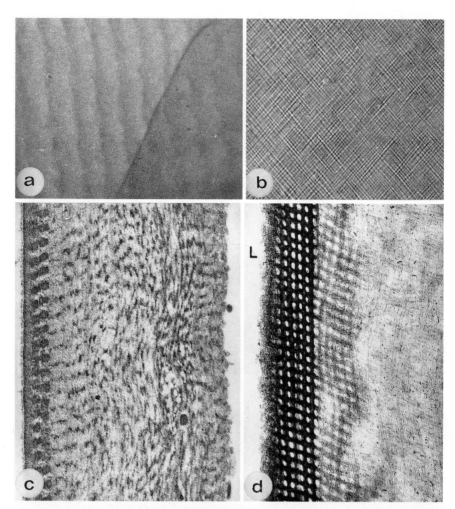

Fig. 42a-d. Larva of *Bibio hortulanus* (Diptera, Bibionidae). After staining with Congo red *a* anisotropic crossbands and *b* a grid texture can be seen in polarized light. *c,d* Electron micrographs show that this is due to an electron-dense layer which is present on the lumen side (*L*) followed by a layer with a less electron-dense grid texture. *d* An oblique section shows the electron-dense grid texture. *a* 120x; *b* 500x; *c,d* 25 000x (After Peters et al. 1978)

⟶

Fig. 43a. Longitudinal section of the multilayered peritrophic membrane of *Penthetria holosericea* (Diptera, Bibionidae); 22 000x. *b* Schematic representation of the gut and cardia of *Miastor metraloas* (Cecidomyidae, Nematocera, Diptera). *BL* Caecum; *BZ* formation zone; *CA* cardia; *ED* hindgut; *K* cuticle; *MD* midgut; *MG* Malpighian tubules; *MS* microvilli; *PM* peritrophic membrane; *OE* esophagus; *VC* valvula cardiaca. *c* Section through a peritrophic membrane of *M. metraloas*; 38 000x (After Peters et al. 1978)

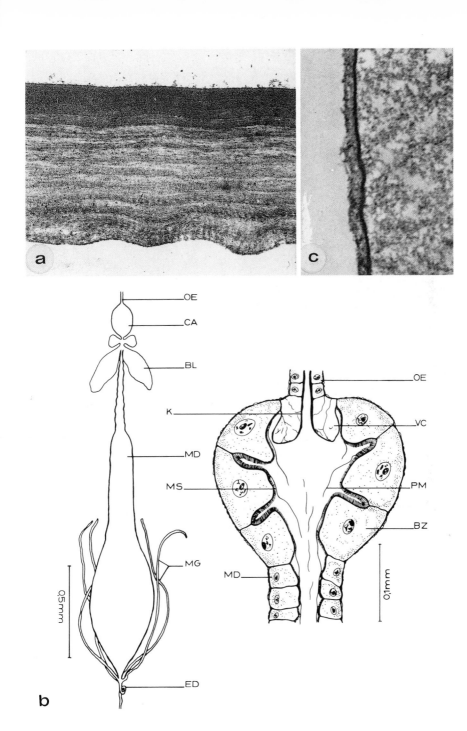

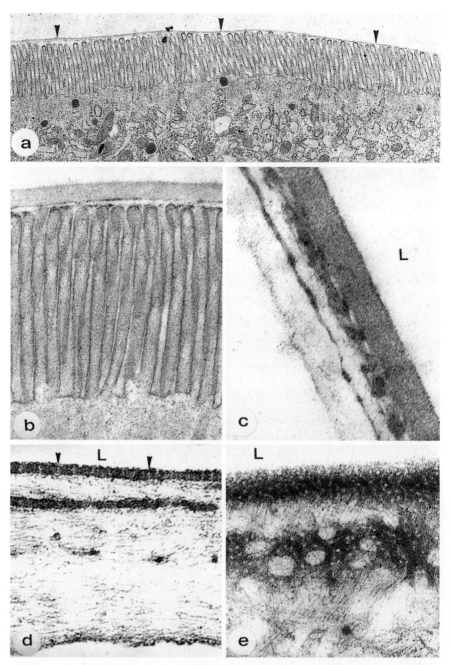

Fig. 44a-d. Larva of *Aedes aegypti. a,b* Formation of the electron-dense layer on the lumen side of the peritrophic membrane. *c* Fully formed peritrophic membrane. *L* Lumen side. *d,e* Larva of *Culex pipiens. d* Longitudinal and *e* tangential section of the peritrophic membrane showing the fine pores (*arrowheads*) in the electron-dense layers of the lumen side (*L*) as well as in the interior of the membrane. In e the grid texture of the microfibrils can be seen. *a* 9000x; *b* 24 100x. *c-e* 62 000x

have microvilli of rather uniform height. Contrary to the findings of Richards and Richards (1971), Peters (1979) found longer microvilli with small or even negligible interstices. One would expect prominent Golgi complexes in these cells, but these were not observed. Richards and Richards emphasized that the number of cells involved in the formation of the peritrophic membrane is uncertain but obviously rather small. It seems that only the anteriormost eight to ten rows of cells are involved in the secretion of the peritrophic membrane. Cross-sections show that each row comprises about 40 cells. Therefore, the peritrophic membrane is produced by approximately 300–400 cells. However, the appearance of these cells does not differ from the posterior rows of cells. Nothing is known about the function of these cells in the posterior part of the cardia. The authors observed secreted material between the microvilli of the first cells. It aggregated over the tips of the microvilli to form a granular-appearing layer, the thickness of which increased continuously. Such material was recognized with certainty at least on top of the third cell. It is reinforced by layers of different electron density. The electron-lucent layers contain microfibrils. Zhuzhikov (1964) could show in shadowed whole mounts of membranes that these are arranged in an orthogonal texture (Zhuzhikov 1964). How these microfibrils are oriented is still unknown. The electron-dense layers consist of or at least contain proteoglycans (Peters 1979). Richards and Richards (1971) called these layers "fibrous layers" as they believed that they contain chitin microfibrils. The electron-dense layers are not continuous layers. In tangential sections they show a system of strands and holes, resembling an irregular honeycomb pattern, whereas in cross-sections electron-dense knots and streaks appear which may be connected by single or double lines of less electron-dense appearance. The holes are irregularly round or elongate with diameters in the range of 60–100 nm, sometimes even 150–180 nm. Another fine, granular layer occurs on the epithelium side of the membrane. First instar larvae fed or unfed have a peritrophic membrane with a thick, granular layer on the lumen side and with a first electron-dense layer. Second instar larvae have a thicker membrane, and third and fourth instar larvae have three to four, and only occasionally up to six electron-dense layers. The fully formed peritrophic membrane of older larvae (Fig. 44) has a thickness in the range of 0.6–0.8 µm with a granular-appearing layer on the lumen side which is 200 nm thick. The production of peritrophic membranes does not depend on the ingestion of food, as newly hatched but unfed larvae start already with its formation. The newly formed peritrophic membrane is already of uniform thickness. This seems to be the result of the rate of secretion, and not due to molding in a "cardia press".

A survey of a dozen genera of mosquitoes not mentioned particularly showed that there are usually three to five interrupted electron-dense layers in their peritrophic membranes (Richards 1975). Only in some specimens of *Culex restuans* and *Anopheles albimanus* areas with 12, 15 or even 20 electron-dense layers were found.

In *Anopheles stephensi* larvae the peritrophic membrane is only slightly different from that of *Aedes aegypti* (Peters 1979). It has a total thickness of about 400–600 nm, a fine granular layer on the lumen side with a width of 80–100 nm, another granular layer on the epithelium side, and four to seven electron-dense layers with irregular holes (see Fig. 71). Histochemical tests showed that granular and electron-dense layers contain proteoglycans.

87

In larvae of *Culex pipiens* the peritrophic membrane differs from that of the preceding species in that on the lumen side it has a layer which does not have a granular appearance but in cross-sections a palisade pattern and in tangential sections an irregular pattern of fine pores with a diameter in the range of 15 nm (Fig. 44e). The spacings between the pores range between 10–20 nm. The whole layer has a thickness of 50–80 nm. It is followed by an electron-lucent layer with microfibrils and a thickness between 50–100 nm. A second electron-dense layer with fine pores follows which differs from the first one in that it, moreover, has irregular holes with diameters ranging up to 100–150 nm. The larger part of this type of peritrophic membrane consists of electron-lucent material with bundles of microfibrils arranged in an orthogonal texture. A third and sometimes a fourth electron-dense layer appear to be more or less incomplete. On the epithelium side of the membrane there is a thin and irregular granular layer. Histochemical tests showed that the electron-dense layers contain proteoglycans. The fully formed peritrophic membrane of this species has a thickness between 1.2–2.0 μm.

Larvae of Chironomidae were investigated comparatively by Platzer-Schultz and Welsch (1969, 1970) and Platzer-Schultz and Reiss (1970; Fig. 45). The cells of the formation zone of *Chironomus strenzkei* and *Chironomus thummi* were investigated electron microscopically, and seven other species were investigated light microscopically. There are four groups of cells: (1) large cells at the anterior end contain large amounts of rough endoplasmic reticulum, free ribosomes, Golgi complexes and electron-dense granules containing secretory materials, but relatively few mitochondria; (2) smaller cells which are rich in mitochondria, and which have deep apical infoldings with microvilli; (3) a few slender cells which seem to have endocrine functions; (4) opposite the posterior part of the valvula cardiaca five rows of small cells are covered by electron-dense material. Most of the peritrophic membrane is secreted by the first cells. It is reinforced by material which is contributed by the cells of the second type. This contains microfibrils which are arranged randomly. The peritrophic membrane of *Chironomus strenzkei* has a total thickness of 0.44 μm. It has on its lumen side a palisade pattern of electron-dense rods which have a length of 110 nm. The peritrophic membrane of *Chironomus thummi* has a thickness between 1.7–1.8 μm; Fig. 45). In this case the electron-dense rods have a length of about 80 nm. Histochemical tests have shown that these rods contain proteoglycans (Peters 1976).

Platzer-Schultz and Welsch (1969) rejected the hypothesis of Mercer and Day (1952), because the palisade pattern seems to be formed without any contact with the surface of the microvilli. The authors did not realize that this hypothesis was defined for the formation of the chitin-containing network of microfibrils, but not for structures which are of different composition and dimensions. The palisade pattern probably results from self-aggregation. The same might apply to similar structures on the lumen side of the peritrophic membranes of the genera *Culex* and *Odagmia*.

In *Odagmia (Simulium) ornata* the peritrophic membrane has a thickness between 1–2 μm. Isolated and cleaned from food it proves to be a straight tube of great elasticity. Again, only a few cells in the anterior part of the cardia are responsible for the formation. The remaining cells of the cardia are even larger than the formation cells. Their function is still an enigma. The formation of the single

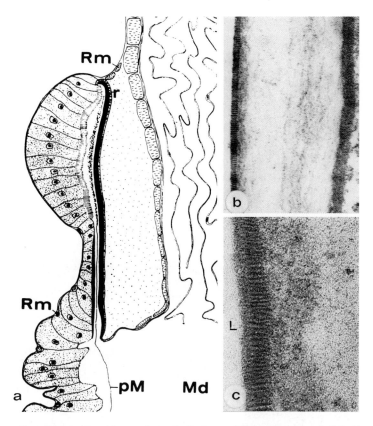

Fig. 45a-c. Formation of the peritrophic membrane in the larva of Chironomidae. *a* Left half of the cardia of *Cryptochironomus* sp. *Md* Midgut; *pM* peritrophic membrane; *r* ring of imaginal cells; *Rm* circular muscles (after Platzer-Schultz and Reiss 1970). *b,c* Electron micrographs of longitudinal sections through the peritrophic membrane of *Chironomus thummi*. Electron-dense rods are present on the lumen side (*L*). *b* 8600x; *c* 27 400x

peritrophic membrane takes place continuously. Fine, granular material appears on the tips of the most anterior cells of the cardia. Slightly more posteriorly the amorphous material transforms into a double layer of rods, probably by self-aggregation. The thin rods of the first layer, which are afterwards on the lumen side, are attached in groups of four to the edges of the more compact rods of the second layer, resembling tuning forks (Fig. 46b). The length of both types of rods could be determined only tentatively as 30 nm. The width of the compact rods was 20 nm. The electron-lucent spacings between "tuning forks" as well as the rods of the second layer were in the range of 10 nm. These values were obtained after standard fixation with glutaraldehyde and osmium as well as after several histochemical reactions. Different values resulted after fixation with permanganate. In this case the compact rods had a diameter of about 25 nm and spacings of about 5 nm. Sometimes the regular pattern is disturbed considerably (Fig. 46). The results of histochemical reactions led to the assumption that the rods consist of or at least contain

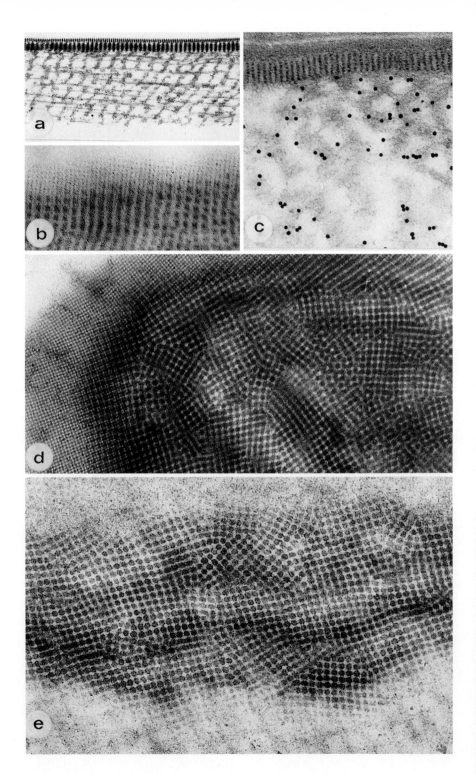

90

proteoglycans. However, there seem to be differences as the layers react differently. The periodic acid-thiocarbohydrazide-silver proteinate reaction introduced by Thiéry (1967) gave a positive reaction with both types of rods (Fig. 46b). But a combination of the proteoglycan reaction of Scott and Dorling (1969) with Thiéry's method reacted only with the compact rods. These differences were also obtained with extraction experiments. If isolated peritrophic membranes of *O. ornata* were extracted with 1 N NaOH at 50 °C for 5 h, only the compact rods of the second layer remained (Fig. 46e). Probably the outer rods consist of or contain mainly glyco-proteins and the inner, compact rods proteoglycans. Similar experiments, supported by biochemical and electron microscopic evidence, were done with the electron-dense layer of the peritrophic membranes of larval and adult blowflies, *Calliphora erythrocephala.*

After both layers with electron-dense rods have been formed in the peritrophic membrane of *Odagmia ornata* they are reinforced by a larger part consisting of an electron-lucent matrix in which bundles of chitin-containing microfibrils are em-bedded. These exhibit a regular orthogonal network (Fig. 46a,e). Their spacings are in the range of 110–130 nm if measured from the middle of a bundle to the next. This is in good agreement with the diameter of the microvilli which is in the range of 85–95 nm. The grid texture of these microfiber bundles is arranged in five to seven layers parallel to the surface of the epithelium. The remarkable ratio of compact rods to bundles of microfibrils is 4:1.

The electron-dense layer on the lumen side of the peritrophic membranes of larvae of Nematocera Culicomorpha appears in different versions:

1.   A continuous layer as in the genera *Aedes* (Fig. 44a-d) and *Anopheles,*
2.   with fine pores as in *Culex* (Fig. 44e),
3.   with fine rods as in Chironomidae (Fig. 45b,c), or
4.   with a double layer of rods with a ratio of 4:1 as in *Odagmia* (Fig. 46).

The larvae of the Brachycera Muscomorpha do not display the structural diversity of the peritrophic membranes of the larvae of Nematocera. They secrete only a single membrane with a simple electron-dense layer on the lumen side. Only its thickness is different: *Drosophila melanogaster* 110–130 nm, *Lucilia sericata* about 250 nm, *Calliphora erythrocephala* 315–500 nm (Figs. 65, 67). The electron-dense layer makes up about 10% of the total thickness of the membrane. Nopanitaya and

---

Fig. 46a-e. Larva of *Odagmia ornata* (Diptera, Simuliidae). *a* Longitudinal section of the peritrophic membrane with two layers of electron-dense rods on the lumen side and the orthogonal texture of microfibrils in the main part of the membrane; 33 000x. *b* Both types of rods give a positive reaction with Thiéry's periodic acid thicarbohydrazide silver-proteinate method showing that carbohydrates are present; 12 000x. *c* Incubation of sections of electron-dense colloidal gold particles with the lectin wheat germ agglutinin reveals the presence of N-acetylglucosamine residues in the grid texture of microfibrils but not in the two layers of rods on the lumen side (see also Sect.5.2); 50 000x. *d* Tangential section of material which had been extracted with phenol shows the intricate orthogonal pattern of both types of electron-dense rods in cross-section; 47 500x. *e* After treatment with NaOH the outer rods have disappeared. The numerical proportion of the two types of rods and the grid texture of microfibrils can be evaluated; 57 000x

Misch (1974) reported that in *Sarcophaga bullata* the thickness of the peritrophic membrane in second and third instar larvae is 142 nm. These authors observed two electron-dense layers on the lumen side with a width of 20 nm separated by an electron-lucent zone. Our own observations in *S. barbata* have shown that in this species the corresponding peritrophic membrane has not only an electron-dense layer on the lumen side, but also a less electron-dense layer on the epithelium side (Nagel and Peters, unpubl. observ.). The formation of the peritrophic membrane of the larva of *Drosophila melanogaster* was investigated light microscopically and described in detail by Rizki (1956).

The adult Brachycera Muscomorpha show a remarkable tendency to increase the number of peritrophic membranes from one to three. This is achieved by folding the originally simple formation zone, as the comparative light microscopic investigations of Zhuzhikov (1963) could show (Fig. 47). During evolution of the system, each new membrane is formed by a new formation zone (Fig. 47). It starts as a very thin, electron-dense membrane (Fig. 50). This is reinforced during later stages of evolution by microfibrillar material until it reaches approximately the thickness of the preceding membrane (Fig. 50). The following examples support this view. The two or three membranes of a given species differ in fine structure and chemical composition (see Sect. 5.2, Figs. 59–62; the dimensions of the different layers are compiled in Table 6).

Suillia flava (Helomyzidae), *Eristalis tenax* (Syrphidae), and the tsetse flies of the genus *Glossina* have a simple formation zone which secretes only a single peritrophic membrane (Fig. 47a). It is characterized by the same electron-dense layer on the lumen side which has been described already in the larvae of the Muscomorpha.

The tsetse flies *Glossina morsitans* and *G. pallidipes* secrete only a single peritrophic membrane as the electron microscopic investigations of Moloo et al. (1970) and Steiger (1973) have shown. It is secreted by three types of cells and consists of an electron-dense layer on the lumen side with a thickness of 15–25 nm and a larger part of low electron density with microfibrils and a thickness between 300–350 nm.

Table 6. Diameter and thickness of peritrophic membranes of flies (Brachycera Muscomorpha) A: adults, L: larva

Species	Stage	Thickness of PM (nm)		
		PM 1	PM 2	PM 3
Eristalis tenax	A		–	–
Glossina morsitans	A	300–350	–	–
Drosophila melanogaster	A	200	45	–
Drosophila melanogaster	L	110–130	–	–
Sarcophaga barbata	A	300–400	50–80	–
Calliphora erythrocephala	A	300–500	60–80	–
Calliphora erythrocephala	L	315–500	–	20–50
Lucilia sericata	A	250–280	120	–
Lucilia sericata	L	250	–	–
Protophormia terrae-novae	A	230–270	160–180	180

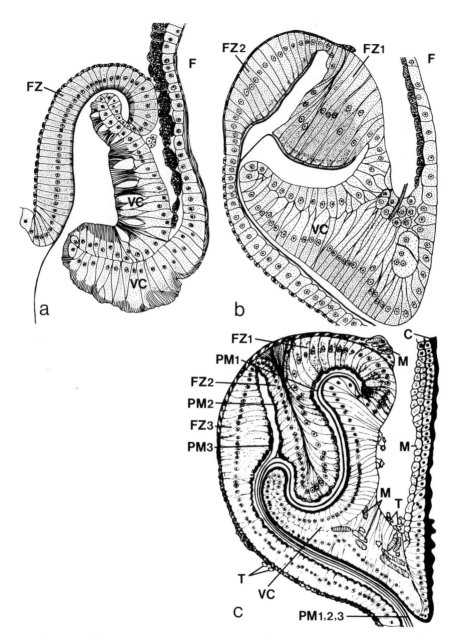

Fig. 47a-c. Semidiagrammatic representation of a median section through the cardia of flies. In *a* and *b* only the left half is drawn. *a Eristalis tenax* (Syrphidae) has only a simple formation zone and therefore secretes only a single peritrophic membrane. *b Musca domestica* shows an upfolding of the formation zone and secretes two peritrophic membranes. *c* The tripartite formation zone of the blowfly *Calliphora erythrocephala* is able to secrete three morphologically different peritrophic membranes. *C* Cuticle; *F* foregut; *FZ 1–3* formation zone 1–3; *M* muscle strands; *PM 1–3* peritrophic membranes 1–3; *T* tracheae; *VC* valvula cardiaca. *a,b* After Zhuzhikov (1963); *c* after Becker (1977)

The newly formed membrane in the cardia is 3–4 µm thick, but when it has left the cardia its thickness is reduced to the above mentioned 300–350 nm. It has not been proved if this is due to the cardia press, as has been assumed by Wigglesworth (1929), or to dehydration of the newly formed membrane. Willett (1966) and Harmsen (1973) were the first to observe that during the first hours after emergence a thin and highly twisted first part of the peritrophic membrane is formed which results in a tight end (Fig. 12). This is of great importance in connection with the uptake of trypanosomes together with the first blood meal (see Sect. 6.6).

In *Drosophila fasciata* Zhuzhikov (1963) found only a single peritrophic membrane. But electron microscopic investigations showed that *D. melanogaster* secretes two membranes (Peters 1976). The inner membrane has a total thickness of 200 nm and an electron-dense layer on the lumen side of 23 nm. The second membrane is electron-dense, very thin with a diameter of 45 nm and very fragile (Fig. 50). King (1988) described in detail the morphology of the cardia and the formation of peritrophic membranes in this species. The border between foregut and midgut epithelium is not on top of the large epithelial fold in the cardia, but about one-fifth down the side of the valvula cardiaca (Fig. 49). King assumed that the electron-dense layer on the lumen side of the peritrophic membrane originates in an unknown way along the surface of the cuticle of the valvula cardiaca. This assumption was made because there seemed to be no traces of secretion by the most anterior midgut cells. However, a small piece of an electron-dense layer appears at the appropriate site in Fig. 48b and is marked as L1. Comparative investigations show that this should be the origin of the electron-dense layer and that there are as yet no hints that the valvula might be involved in the formation of this layer as King suggested. The origin of the second and largest layer is easily visible even with the light microscope (Fig. 48). The third "layer" is made up of fine, electron-dense granules only. The "fourth layer" as King called it, is in fact the very thin, second or outer peritrophic membrane (PM 2). It is secreted by a large group of well-defined cells (6 in Fig. 48). King's observations show that in *Drosophila* there is the beginning of the evolution of two peritrophic membranes due to the presence of two specialized formation zones. However, this is not as clear-cut as in the following species.

In Muscidae, Sarcophagidae, Tachinidae and some smaller families Zhuzhikov (1963) reported two peritrophic membranes which are secreted by two formation zones (Fig. 47). In *Musca domestica* these membranes are rather similar.

The stablefly, Stomoxys calcitrans, according to Lehane (1976), forms a single peritrophic membrane with five structurally or histochemically different layers. However, it seems that there are in fact two membranes which were only closely apposed in the preparation. The description allows the suggestion that the layers 1 and 2 correspond to the first or inner peritrophic membrane (PM 1), and that the layers 3–5 are equivalent to the second membrane (PM 2) of other species, as there are two formation zones. Layer 1 is electron-dense. It has on its lumen side a 3-nm-thick boundary of higher electron density. Pores with a diameter of 3 nm extend through the whole layer. Layer 2 is electron-lucent. Lehane suggested on the basis of histochemical evidence that it probably contains chitin, although tangential sections did not reveal any microfibrils. The thickness of the layers 1–5 varies considerably between 130–400 nm. The electron-dense layer (1) had a thickness of 20–35 nm,

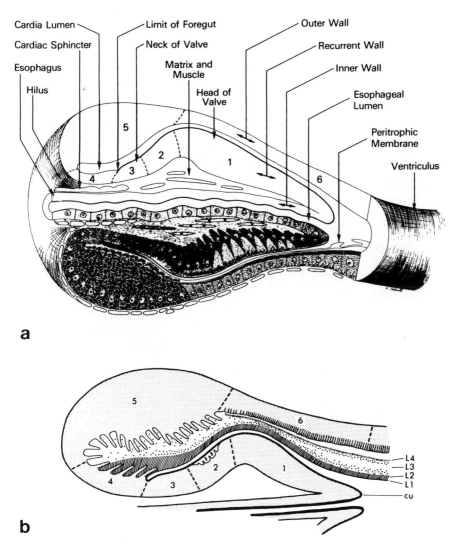

Cardia Lumen
Cardiac Sphincter
Esophagus
Hilus

Limit of Foregut
Neck of Valve
Matrix and
Muscle
Head of
Valve

Outer Wall
Recurrent Wall
Inner Wall
Esophageal
Lumen
Peritrophic
Membrane
Ventriculus

5
2
3
4
1
6

a

5
6
4
3
2
1
L4
L3
L2
L1
cu

b

Fig. 48a,b. Diagrammatic representation of the cardia of *Drosophila melanogaster. a* Morphology. *1–6* Epithelial regions. The inner peritrophic membrane is thrown into folds, whereas the outer membrane is rather straight. *b* Formation of two peritrophic membranes in the different regions of the cardia (see text). *L1–2* represent the layers of the inner peritrophic membrane (PM 1). L4 is formed by region 6 and corresponds to the thin outer PM 2. *cu* Cuticle of the valvula cardiaca (foregut). Not drawn to scale (After King 1988)

and the electron-lucent second layer 20–160 nm; layer 3: 20–70 nm, layer 4: 10–60 nm, and layer 5: 20–70 nm.

The fleshfly, *Sarcophaga bullata*, was investigated electron microscopically by Nopanitaya and Misch (1974). The authors found obviously only the highly convoluted inner peritrophic membrane with its characteristic electron-dense layer on

Fig. 49. Scanning electron micrograph of the cardia of the blowfly, *Calliphora erythrocephala*. *CA* Cardia; *MG* midgut; *T* trachea; 520x

the lumen side and a total thickness of 250 nm. Zhuzhikov (1963) reported from light microscopic studies that Sarcophagidae have two peritrophic membranes. This could be confirmed by electron microscopic investigations which show that in *Sarcophaga barbata* two peritrophic membranes are formed by the bipartite formation zone (Nagel and Peters, unpubl. observ.). The inner membrane (PM 1) has a thickness between 300–400 nm, a conspicuous electron-dense layer on the lumen side and fine fringes on the epithelium side. The outer membrane (PM 2) is much thinner. It has a thickness between 50–80 nm. It consists of an electron-lucent matrix with a random texture of microfibrils in it. Thus, it is similar in appearance to the matrix of PM 1. Electron-dense particles and fine fringes are distributed on both sides of the membrane (Fig. 61).

In adult Calliphoridae Zhuzhikov (1963) found two peritrophic membranes in *Pollenia rudis, P. vespillo, P. varia, Lucilia sericata, L. silvarum* and *L. caesar*, and three membranes in *Calliphora erythrocephala, C. vomitoria, C. uralensis,* and *Protophormia terraenovae*. However, our own electron microscopic investigations have shown that in *Lucilia sericata* three membranes are formed. The inner peritrophic

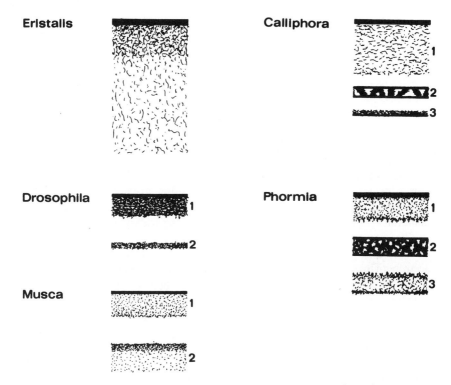

Fig. 50. Diagrammatic survey of the structure and thickness of the peritrophic membranes of adult flies. Drawn to scale

membrane (PM 1) has a thickness between 250–280 nm and an electron-dense layer on the lumen side with a thickness of 25–28 nm. PM 2 is 120 nm and PM 3 only 60 nm thick.

In the blowfly *Calliphora erythrocephala* Becker (1977) has given the most detailed description of the formation of these membranes (Fig. 47). The first cells of the formation zone 1 (FZ 1) are characterized by a large amount of rough endoplasmic reticulum and small stacks of Golgi cisternae. The apical parts of these cells are crowded with vesicles which have a diameter of up to 300 nm. Their contents are more or less electron-dense. These cells have slender microvilli which are 3–4 μm long. Their secretion products appear either as small vesicles with a diameter of 300 nm or as lamellate structures with a length ranging between 0.2–2.0 μm. The latter consist of an electron-dense, 7–8-nm-thick layer and a less electron-dense material apposed to it as a 20-nm-thick layer. These lamellae appear to be very flat vesicles as sometimes the ends are connected and inflated. These structures are remarkable as they are the precursors of the electron-dense layer on the lumen side of the inner peritrophic membrane (PM 1). They can be compared with a switching yard where the cars are linked to a train. The group of vesicles and lamellated structures is converted into the continuous electron-dense layer which has a thickness of about 25 nm. The following cells of the formation zone secrete less electron-dense material thus reinforcing the electron-dense layer. This matrix material contains microfibrils arranged randomly. At first it has an irregular border facing the cells and appears to be loosely packed. In the middle of the formation zone it has a regular thickness of 0.8–1.0 µm. In the posterior part of the cardia it seems to be more compact and attains its definite thickness of 300–500 nm. The cells of the second formation zone (FZ 2 in Fig. 47) are only half as high as those of formation zone 1 (FZ 1). They contain more stacks of Golgi cysternae and vesicles as compared with those of FZ 1. Vesicles with a diameter of up to 1 µm contain electron-lucent, fine, filamentous material but rarely electron-dense material. The vesicles may fuse to form very large vesicles in the apical parts of these cells. The secreted material forms at first a continuous strand on top of the microvilli and then is transformed into a double layer with connecting struts. This is spun off the wedge-shaped end of formation zone 2 (FZ 2 in Fig. 47). The second peritrophic membrane (PM 2) has a thickness of 80–100 nm. Zhuzhikov (1963), Smith (1968) and Zimmermann et al. (1969) assumed that the third peritrophic membrane of the blowfly, *Calliphora erythrocephala*, is formed by the midgut epithelium. However, electron microscopic investigations revealed that it is formed by an additional third cushion of cells of the formation zone in the cardia and not by cells of the anterior midgut (DePriester 1971; Becker et al. 1975; Peters 1976) (Fig. 47). It is composed of long slender cells with microvilli which are 2–3 µm long. The cells resemble those of formation zone 1, but they have less Golgi stacks. Two types of vesicles with diameters of up to 300 nm are present which have either electron-lucent or to a lesser extent electron-dense contents. No secretion products were observed in the first third of the third formation zone (FZ 3 in Fig. 47). But on top of the microvilli of the following cells, a thin layer can be seen which attains at the end of formation zone 3 a thickness of only 30–50 nm. It is not astonishing that such a thin and fragile membrane has been missed in light microscopic investigations. Moreover, it is impossible to consider these third peritrophic membranes in biochemical and physiological investigations. The three types of peritrophic membranes are formed continuously, leave the cardia and form an extensive accumulation of membranes in the adjacent part of the midgut. Such a reserve seems to be useful in case of a sudden income of larger amounts of food which otherwise might draw the newly formed membranes out of the cardia and thus disturb the formation of new membranes. The importance of the valvula cardiaca for the formation of intact peritrophic membranes has been more a matter of speculation than investigation. Those parts of the valvula which lie opposite the formation zone 1 have a complex system of infoldings. These remind one of a transport epithelium, however, the characteristic accumulation of mitochondria is lacking. Dark cells or groups of cells are present, the cytoplasm and organelles of which appear to be condensed, and with inflated infoldings. Other cells have rather narrow infoldings. The significance of these structures is unknown. Perhaps they are of importance for the unknown processing and dehydration of the newly formed membrane.

The convoluted masses of peritrophic membranes in the midgut favor misinterpretations of their sequence. Thus, Smith (1968) called the inner peritrophic membrane of *Calliphora erythrocephala* PM 2. In order to number the membranes they have to be followed down the respective formation zones to the posterior part of the cardia.

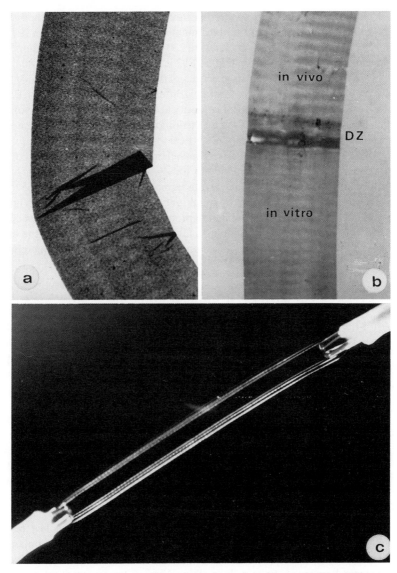

Fig. 51a-c. Anisotropic crossbands were observed in the peritrophic membranes of Diptera, in this case in the inner peritrophic membrane (PM 1) of adults *Calliphora erythrocephala, a* in autoradiographs after the incorporation of ^{35}S-methionine; *b* after staining with the dichroitic dye Congo red; and *c* in an untreated membrane tube extended between capillaries using normal light at an. acute angle near to total reflection. The band pattern can only be seen in the periphery. In these examples only macrobands are shown. The regular pattern is disturbed if the flies are treated with a temperature shock or isolated cardiae are transferred to an artificial medium. Under these circumstances a disturbed zone (*DZ*) is formed which can be used as a marker for the beginning of the respective experiment. *a* 43x (after Zimmermann et al. 1973). *b* 43x (after Becker et al. 1975). *c* Light micrograph (Courtesy of Prof. Dr. Zimmermann, Würzburg)

In *Phormia terrae-novae* three peritrophic membranes are formed as in *Calliphora erythrocephala*, but the third membrane is much thicker than in the latter species (Fig. 57). PM 1 has a thickness of 230–270 nm and an electron-dense layer on the lumen side which is 40 nm thick. PM 2 is 160–180 nm and PM 3 180 nm thick.

The most remarkable characteristics of the peritrophic membranes of blowflies are their morphological and biochemical differences and asymmetries (see Sect. 5.2) which should have physiological consequences. The electron-dense layer which is rich in proteoglycans lies in PM 1 on the lumen side, but in PM 3 on the epithelium side; PM 2 appears to be more symmetrical.

Finally, it should be emphasized that all these investigations do not concern the problems of regulation of formation by muscles and nerves in the cardia. Nothing is known about the function of the valvula cardiaca epithelium opposite the formation zone with its elaborate system of membrane foldings. Moreover, we do not understand the function of a large number of prominent and very tall columnar cells in the formation zone as it seems that only a few preceding smaller cells at the beginning of the formation zone are able to secrete the respective peritrophic membrane.

3.3.4 Crossbands in Peritrophic Membranes

After the injection of glucose-^{14}C into blowflies, *Calliphora erythrocephala* (Diptera), Zimmermann et al. (1969) did not find an even distribution of label in autoradiographs of peritrophic membranes as expected, but instead a periodic incorporation. The mean width of the labeled and unlabeled crossbands was in the range of 40 μm.

The same result was obtained by the injection of ^{35}S-methionin and ^{35}S-cysteine (Zimmermann et al. 1973; Fig. 51a). Further investigations showed that similar crossbands can be observed in polarized light after staining of peritrophic membranes with the dichroitic dye Congo red (Peters et al. 1973; Fig. 51b) and in untreated membranes if normal light was applied near total reflection (Zimmermann, pers. comm.; Fig. 51c). In this way, comparative results could be obtained with independent methods which supports the view that these crossbands cannot be artifacts.

Zhuzhikov (1966) described crossband patterns in only a few peritrophic membranes of adult houseflies, *Musca domestica*, after observation in polarized light and staining with Congo red. However, most of the membranes of this species appeared homogeneous as well as all membranes from adults and larvae of *Calliphora erythrocephala* and *Aedes aegypti*. These negative results may be due to the salt content of the staining solution and the short time of staining used by Zhuzhikov.

Peters et al. (1973) found macrobands with a width between 50–80 μm in the peritrophic membranes of several species (Table 7). If membranes stained with Congo red are rotated between crossed polars the extinction is shifted every 90° from a band into an interval and vice versa. Therefore, periodic units = band + interval were measured (Table 7). The results indicate that two anisotropic sections are present which are arranged periodically in series. The oriented structures must be oriented perpendicular to each other.

Table 7. Crossbands in peritrophic membranes of Diptera (Peters et al. 1973)

Species	Stage/sex[a]	Method[b]	n[c]	Periodic unit[d]	Diameter[e]	PM[f]
Calliphora erythrocephala,	adult	Au gluc	34	88 (15)	551	
Calliphora erythrocephala,	adult	Au met	23	82 (19)	555	
Calliphora erythrocephala,	M+F	PL	28	91 (21)	496	I+O
Calliphora erythrocephala,	M	PL	8	90 (17)	503	I
Calliphora erythrocephala,	F	PL	38	79 (23)	435	I
Calliphora erythrocephala,	F	PL	24	72 (24)	451	O
Calliphora erythrocephala,	larva	Au gluc	3	94 (9)	420	
Calliphora erythrocephala,	larva	Au met	7	68 (16)	505	
Calliphora erythrocephala,	larva	PL	58	52 (14)	521	
Lucilia sericata,	adult	Au met	3	40 (3)	365	
Lucilia sericata,	M+F	PL	17	50 (11)	344	I
Lucilia sericata,	M+F	PL	10	54 (10)	350	O
Lucilia sericata,	larva	PL	16	61 (12)	350	
Musca domestica,	M+F	PL	4	63 (15)	512	I
Musca domestica,	M+F	PL	4	26 (2)	420	O
Musca domestica,	larva	PL	10	42 (17)	350	

[a]Sex: M= male, F= female.
[b]Observed in Au= autoradiographs; PL= polarized light after staining with Congo red.
Labeled compound: gluc= ^{14}C-glucose; met=^{35}S-methionin.
[c]n= Number of experiments.
[d]Periodic unit = band + interval in µm; the standard deviation is given in parentheses.
[e]Diameter: average diameter of peritrophic membranes in µm.
[f]PM: I= inner, O= outer.

Moreover, microbands with a width of 10–14 µm were observed in peritrophic membranes of adult blowflies, *Calliphora erythrocephala* and *Lucilia sericata*, but in the housefly, *Musca domestica* only in the inner membrane (PM 1). In adult *Musca domestica* a large difference was observed between the width of the macrobands in the inner and outer peritrophic membrane. Such differences have not been found in males and females of *Calliphora erythrocephala*. The angle of extinction between macro- and microbands amounts to 45°. Microbands appear more regularly if the membranes are stained for more than 48 h with desalted solutions of Congo red. The dye may contain up to 22% sodium by weight. Microbands seem to be absent in larval membranes. This may be due to the smaller width of the macrobands in larval membranes which could result in microbands with a width below the resolution of the polarized light microscope.

Agreement between the results obtained by autoradiography and measurements of macrobands in polarized light is satisfactory with the exception of larvae of *Calliphora erythrocephala* (Table 7).

At first sight there seems to be a linear relationship between the width of the periodic units and the diameter of the peritrophic membrane tube. However, the correlation factors are only of the order of 0.83. Therefore, we cannot decide whether the relationship between these two parameters is really linear.

101

During cultivation of cardiae in vitro, the width of the bands is reduced in the newly formed membranes if the conditions in the medium are adverse (Sect. 3.3.7). At best, the width of bands of membranes formed in vivo is obtained.

Macro- and microbands seem to be the result of interference of double-diffracted beams with hitherto unknown structural elements. This view is supported by the results obtained with peritrophic membranes of larvae of the blackfly, *Odagmia* (*Simulium*) *ornata* (Diptera). In this case three orthogonal lattices of different dimensions are present (Sect. 3.3.3.2 and Fig. 46). The Moiré pattern, which can be observed in polarized light after staining of the membranes with Congo red, is highly complicated and corresponds to electron micrographs of tangential sections or negative staining preparations. Disturbances in the orthogonal arrangement of the two layers of rods on the lumen side of the membrane are obviously reflected in the Moiré pattern.

If proteins and proteoglycans are extracted from the peritrophic membranes of blowflies, the periodic bands still remain. Therefore, the corresponding ultrastructure might be either identical to or firmly attached to chitin structures.

3.3.5 Induction of Formation of Peritrophic Membranes

In most species the formation of peritrophic membranes is a continuous process which starts after emergence and which is independent of the ingestion of food. Georgi (1969b) obtained membranes from crayfish, *Orconectes (Cambarus) affinis* (Crustacea, Decapoda) which had not been fed for several months.

However, in some groups membrane formation is induced by ingested food. Reeve et al. (1975) reported that a species of the Chaetognath, *Sagitta hispida*, surrounds its prey with peritrophic membranes. Three to four h after feeding elongate fecal pellets are voided.

The induction of the formation of peritrophic membranes has been investigated mainly in Nematocera, and is of medical importance, especially in Culicidae. Yaguzhinskaya (1940) was probably the first to observe that the ingestion of blood triggers the formation of peritrophic membranes in female mosquitoes. The first appearance of a distinct peritrophic envelope seems to be specific for the different species which have been investigated so far, but there are discrepancies between the results of the authors. According to Freyvogel and Stäubli (1965), a peritrophic envelope is observed in *Aedes aegypti* 5–8 h, in *Anopheles gambiae* 13 h, and in *Anopheles stephensi* 32 h after feeding. This phenomenon is of great importance for the development of *Plasmodium* species which will be discussed in Section 6.6.

The prerequisites for the formation of peritrophic membranes have been determined for *Aedes aegypti*, *Anopheles stephensi* and *A. gambiae* by Freyvogel and Jaquet (1965). In *Aedes aegypti* there is a correlation between the amount of blood ingested and the amount of secreted membranes, whereas in *Anopheles stephensi* no such correlation seems to exist and secretion obeys the all or nothing principle. The administration of blood, serum, water or even air resulted in the formation of peritrophic membranes which gave a positive chitosan reaction. Even the injection of physiological saline into the rectum induced the formation of peritrophic membranes.

The enema technique was also used by Berner et al. (1983) in *Anopheles stephensi*. An enema with salt solution triggered the secretion of granules which were already present in the cells of the stomach. Twenty hours later a more or less differentiated peritrophic membrane could be observed by electron microscopy. As there was no blood bolus in the midgut, the peritrophic membranes floated freely in the salt solution. For comparison, compact peritrophic membranes appeared in the posterior midgut 30–72 h after a blood meal (Freyvogel and Stäubli 1965). After an enema with a salt solution the gut volume was reduced rapidly. This could have been achieved by rapid water transport through the midgut epithelium into the hemolymph. The resulting increase in the salt concentration could trigger the secretion of peritrophic membrane material and, therefore, the formation of peritrophic membranes (Berner et al. 1983).

Experimental distension of the stomach by an enema with salt solution did not increase proteolytic activity in the gut. This is in agreement with the finding of Briegel and Lea (1975) that in *Aedes aegypti* the uptake of soluble proteins triggers the release of proteases from the midgut epithelium.

In *Aedes aegypti* two blood meals separated by an interval of 10 h led to the formation of two peritrophic envelopes (Freyvogel and Stäubli 1965). The assumption that male and female Culicidae do not form peritrophic membranes after the ingestion of nectar should be reexamined.

3.3.6 Rate of Formation of Peritrophic Membranes

Waterhouse (1954) pointed out already that there is no correlation between the rate in which food passes down the alimentary tract and the rate of production of peritrophic membranes. In larvae of *Eristalis tenax* (Diptera, Syrphidae) he could determine the rate in which food passes down the gut by marking dilute and filtered liquid manure with finely powdered animal charcoal or Prussian blue. At 30 °C the food passage amounted to 50–75 mm/h. Under the same conditions 6 mm of peritrophic membrane was formed; one larva secreted up to 18 mm membrane per hour. In water 5 mm membrane was formed at 30 °C. In starving larvae the rate of formation dropped to about 2 mm membrane per hour. Aubertot (1932) observed in the same species a production of 6.1 mm membrane. In blowfly larvae, *Calliphora fallax* and *C. augur*, which had been removed from their food and kept at 30 °C, a continuous peritrophic membrane was produced at a rate between 5–10 mm/h for the first 24 h; food passed down the alimentary canal at about 75 mm/h (Waterhouse 1954).

The rate of formation was observed mainly in species or stages of insects which live in water and extrude the peritrophic membrane from the anus as an unbroken tube. Sometimes it attains considerable length in undisturbed animals, e.g. in mosquitoes or blowfly larvae. However, mosquito larvae may bend around and ingest the protruding membrane tube (Richards and Richards 1971; for Crustacea, see also Sect. 6.7). In flies the rectal papillae are covered with spicules which are believed to assist in dragging the peritrophic membranes posteriorly and to eventually tear the membrane (Wigglesworth 1972; see Chap. 4). Frequently, the feces are solid or semisolid. In the grasshopper *Locusta migratoria* and the cockroach *Periplaneta americana*, the peritrophic membranes which originally surround the food residues in the midgut are fragmented during formation of the fecal pellets. Generally, the fate of peritrophic membranes in the posterior parts of the gut is unknown in other groups of animals.

Despite these difficulties, an attempt was made to estimate the rate of formation of peritrophic membranes in some insects from the discharge of fecal pellets. In the larva of *Aeschna* sp., even without feeding, about two peritrophic sacs may be discharged each day (Aubertot 1932a,b). Nymphs of several species of Odonata Zygoptera produced peritrophic membranes at a rate of three per day if they were fed, whereas hungry animals voided only one per day (Waterhouse 1954).

Bernays (1981) reported that in actively feeding locusts, *Schistocerca gregaria*, a peritrophic membrane is delaminated from the midgut epithelium, including the caeca, at intervals of approximately 15 min. Waterhouse (1954) determined the rate of formation of peritrophic membranes in an earwig, *Labidura truncata* (Insecta, Dermaptera). Well-nourished animals of this species produced 1.6 mm of peritrophic membranes per hour at 30 °C, and 0.25 mm PM at 18 °C; starving animals secreted 0.6 mm PM per hour at 30 °C. Peters et al. (1979) investigated the common earwig, *Forficula auricularia*, and reported a formation rate of membranes in vitro in *Locusta* hemolymph of 0.5–0.65 mm peritrophic membrane per hour at 25 °C. In a weevil, *Rhynchaenus fagi* (Coleoptera, Curculionidae), peritrophic membranes are used for cocoon formation at the end of the larval period (Sect. 6.8). Streng (1973) observed at this stage a production of 10 mm membranes per minute! In larvae of wasps and bees the midgut is closed at the posterior end. Therefore, the peritrophic membranes accumulate in the posterior part of the midgut. Rengel (1903) maintained that at least six membranes are secreted each day.

More reliable data could be obtained in Diptera with their continuous production of tube-like peritrophic membranes. In the tsetse fly, *Glossina pallidipes*, membrane formation starts immediately after emergence. Harmsen (1973) dissected the membrane from the gut at periods up to 80 h (Fig. 13). Measurement of these membranes showed a linear relationship between length and time for the first 30 h and a formation rate of approximately 1 mm membrane per hour. Subsequently, a gradually declining rate was observed until the membrane reached a length of 60–80 mm at 80 h. The growth rate was slightly higher at 25 °C than at 30 °C which indicates that 30 °C might be above the optimal temperature for this species. Harmsen emphasized that there is no indication of a rapid increase of formation after a blood meal, as susggested by Wigglesworth (1929), or that shortly after feeding secretion is higher in young than in older flies, as assumed by Willett (1966).

The peritrophic membranes of blowflies could be labeled by two methods, either radioactively or by temperature shock. In adult *Calliphora erythrocephala*, the formation rate was determined by autoradiography after labeling with ^{14}C-glucose (Zimmermann et al. 1969) to be 3.5 mm ± 1.0 mm membrane per hour at 20 °C. Investigations on the formation of peritrophic membranes in vitro showed that dissection and transfer of the cardia into a new medium resulted in a disturbance of the regular pattern of bands in the membrane during the first 20 min in the new medium or at temperatures below 5 °C (Zimmermann et al. 1973; Becker et al. 1975, 1976). A thickened zone lacking the regular pattern of anisotropic crossbands was called the "disturbed zone". Its width varied from 0.2–0.6 mm depending apparently on the temperature of the medium. It marked the beginning of in vitro production of membranes. Therefore, it was possible to determine quite easily the formation rate of peritrophic membranes by labeling the membranes with low temperature shocks (Becker 1978a). A first temperature shock was applied at 4 °C for 1 h; this resulted in a first disturbed zone in the newly formed peritrophic membrane. Then the insects were transferred for 1–4 h to the temperature at which the rate of formation had to be determined. A second temperature shock was given at 4 °C for 1 h which resulted in another disturbed zone. Each experiment was terminated after a resting period of 1 h at 20 °C. Then the membranes were dissected, stained with a 0.5% aqueous solution of the dichroitic dye Congo red for 24–28 h, spread on slides and observed under normal and polarized light. The distance between the two disturbed zones was measured. It corresponds to the formation rate of peritrophic membrane in the respective period at a given temperature. The results are listed in Table 8. In *Calliphora erythrocephala* the formation rate is apparently slightly higher in female than in male flies (Table 8). The results obtained with this method are in the same range as those determined by Zimmermann et al. (1969, 1973) by autoradiography after the incorporation of labeled glucose and methionine.

In larvae the production increases with the size of the larvae. In third instar larvae the rate of formation decreases as they approach pupation. Waterhouse (1954) determined the formation rate of peritrophic membranes in larvae of *Calliphora augur* and *C. fallax* by measuring the length of membrane protruding from the anus. He found a rate of 5–10 mm membrane per hour at 30 °C. Such large variations have not been found by Zimmermann et al. (1973) and Becker (1978a). However, Waterhouse did not regard the usual accumulations and coilings of peritrophic membranes in the gut. We have never observed a more than a tenfold individual variation as reported by Waterhouse (1954). In a given species large flies have a

Table 8. Formation rates of peritrophic membranes in flies[a] (Becker 1978a)

Species	Number of PM	Rate of Formation (mm/h)		Diameter of PM (μm)
		20 °C	30 °C	
Calliphora erythrocephala, male	3	4.0 (0.7)	6.3 (0.8)	512 (48)
Calliphora erythrocephala, female	3	4.6 (0.9)	7.1 (1.1)	528 (39)
Calliphora erythrocephala, L2	1	2.3 (0.4)	2.9 (0.3)	331 (48)
Calliphora erythrocephala, L3	1	3.6 (0.5)	5.5 (0.6)	525 (66)
Protophormia terrae-novae, adult	3	4.7 (0.7)	6.1 (0.6)	451 (41)
Protophormia terrae-novae, L3	1	3.3 (0.4)	5.4 (0.9)	405 (52)
Sarcophaga barbata, adult	2	3.0 (0.5)	5.6 (0.3)	742 (70)
Sarcophaga barbata, L3	1	5.1 (1.0)	7.2 (1.1)	466 (43)
Musca domestica, adult	2	2.2 (0.2)	3.3 (0.4)	293 (53)
Musca domestica, L3	1	2.7 (0.7)	3.8 (0.3)	265 (39)

[a]The measurements are mean values from ten experiments. Standard deviation is given in parentheses. L2, second instar larva; L3 third instar larva.

higher rate of synthesis than smaller ones. In the large fleshfly, *Sarcophaga barbata*, the formation rate is relatively low compared with smaller species investigated.

In adults and larvae of the blowfly *Calliphora erythrocephala*, Becker (1978a) observed a nearly linear increase in the formation rate up to 20 °C in normally fed flies (Fig. 52) and a further increase in the range of 20–30 °C. In adults the formation rate decreased above 35 °C, in larvae already above 30 °C (Fig. 52b). In starving flies the formation rate is lower by up to 2 mm/h as compared with fed flies. In vivo the maximum rate was obtained with fed as well as starving flies at 35 °C, but in vitro it was between 26–28 °C (Becker et al. 1975). At a temperature of 26–28 °C 6.2 mm of peritrophic membranes per hour were formed in vivo, but only 3.6 mm in vitro. The fine structure of membranes grown in vitro did not differ from those grown in vivo (Becker et al. 1976). The reduced rate of formation observed in vitro may be due to a lack of nutritive substances or hormonal stimulation, or to physiological differences between in vivo and in vitro, e.g. salt and nutrient concentrations, oxygen tension, accumulation of waste products etc.

3.3.7 Formation of Peritrophic Membranes in Vitro

This has been achieved as yet only with the cardia of adult blowflies, *Calliphora erythrocephala*. Just a few preliminary results were obtained with the cardia of the common earwig, *Forficula auricularia* (Peters et al. 1979). Shortly after the isolated cardiae of the earwig had been transferred to the medium, they resumed their regular contractions; however, these ended after at least 15 min. In Leloup's medium up to 3 mm of peritrophic membranes was formed over 15 h. In Hoyle's *Locusta* Ringer with the addition of *Locusta* hemolymph, up to 4.8 mm of peritrophic membranes was secreted after 8 h. The best results were obtained using *Locusta* hemolymph as incubation medium. In 8 h, 4.0–5.2 mm of peritrophic membranes was formed, which means a formation rate of 0.5–0.65 mm of peritrophic membranes per hour at 25 °C.

For the first time Zimmermann et al. (1973) could demonstrate the synthesis of peritrophic membranes in vitro by cardiae of adult blowflies which were incubated in Tyrode solution to which glutamin had been added, and which was saturated with oxygen. However, the quality of such peritrophic membranes was poor. The membranes were rather sticky. In polarized light a reduction of the macroperiods was observed, whereas the microperiods disappeared in most cases. Electron micrographs (Zimmermann et al. 1975) showed that the inner membrane (PM 1) was rather fluffy. Its thickness increased, varying between 450–1200 nm. A conspicuous degeneration of the electron-dense layer of this membrane occurred.

→

Fig. 52a,b. The rate of formation of peritrophic membranes of larval and adult blowflies, *Calliphora erythrocephala*, is plotted against increasing temperature. *a* It is higher in actively feeding than in starving flies. •-• Normally fed flies; °-° starving flies. *b* Adults tolerate higher temperatures than larvae. •-• Adult blowflies; °-° larvae (L3). Each *point* represents the mean of 10 experiments. The *bars* represent confidence limits of 95% (After Becker 1978a)

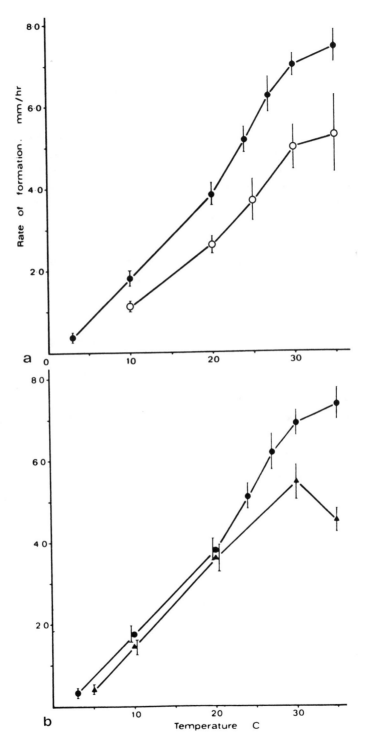

107

Normally, the thin layer passed into a folded part with several layers. In membranes grown in vitro it was sometimes absent or transformed into circles. But sometimes large stacks of electron-dense layers were observed which in one case reached a thickness of 1–6 μm. In the matrix of the chitin-containing part of the inner peritrophic membrane, cellular debris occurred. This was observed in membranes grown in vivo, too, but it disappeared in this case immediately in the formation zone. The second and third peritrophic membrane grown in vitro also increased in thickness. The thickness of the latter varied between 170–770 nm.

More systematic investigations regarding the influence of chemical composition, osmolarity, pH, and temperature of the medium were carried out subsequently by Becker et al. (1975, 1976). Leloup's medium (Leloup 1964) was used as a basic medium as it proved to be superior to Tyrode solution as well as the more complex media used for insect tissue culture, e.g. Grace's or Schneider's medium. The addition of chick embryo extract to the medium apparently had no effect on the rate of formation, whereas the addition of fetal calf serum had deleterious effects. The addition of fetal calf serum or *Calliphora* hemolymph to the culture medium resulted in a reduction of the formation rate and the appearance of very small or irregular crossbands in the peritrophic membranes formed in vitro. *Calliphora* hemolymph had to be heated for 5 min at 60 °C to destroy the phenoloxidase system (Wyatt 1956). Leloup's medium contained per 100 ml: 650 mg NaCl, 25 mg KCl, 30 mg MgSO$_4$ × 7 H$_2$O, 35 mg CaCl$_2$, 80 mg trehalose, 70 mg glucose, 50 mg maleic acid, 25 mg ketoglutaric acid, 5 mg succinic acid, 5 mg fumaric acid, 100 mg yeast extract, 1000 mg lactalbumine hydrolysate, 5 mg streptomycin to inhibit the growth of bacteria; the pH was adjusted to 7.0–7.1 with 1 N NaOH.

Cardiae of flies 2–4 days after emergence were dissected in culture medium, carefully avoiding any strain. About 2 mm of the crop duct and the midgut were left on the cardia. After several rinses in fresh culture medium the cardiae were transferred to sterilized vessels with culture medium which had been sterilized by filtration. The culture medium was saturated continuously with oxygen and kept at a constant temperature by a water bath. The medium was changed every 2 h.

For the evaluation of the newly formed peritrophic membranes the following criteria could be used: (1) rate of formation; (2) fine structure; and (3) width of the anisotropic crossbands. Cardia, midgut and crop duct exhibited irregular convulsive contractions at the beginning of incubation. But after about 15 min regular contractions and a continuous production of peritrophic membranes could be observed. After isolation of the peritrophic membranes the start of formation in vitro could be easily found by observation of the membranes in normal light near the angle of total refraction, or in polarized light after staining with Congo red. The start of in vitro formation was marked by the irregular "disturbed zone", the length of which was quite variable (Fig. 51b). Its length seems to depend on the quality of the medium, and varied between 0.2–0.6 mm of peritrophic membranes per hour. In Leloup's medium the formation rate at 25 °C was 3.0 ± 0.6 mm PM per hour during the first 10 h. Despite sterilization and the addition of antibiotics, a continuous increase in bacteria was found in sections of cardiae which had been incubated for 24–72 h. After 24 h of incubation about 50 mm of peritrophic membranes had been secreted. Afterwards the contractions of the cardia continued, but the formation of membranes ceased.

The osmolarity of the culture medium was varied by the addition of glucose, sucrose, or sorbitol to the basic medium. Osmotic values below that of the basic medium were obtained by reducing the amount of sodium chloride, glucose or trehalose. The pH was adjusted to 6.8, and the temperature of the incubation medium was maintained at 27 ÉC. The values plotted in Fig. 53 correspond to the average rate of peritrophic membrane formation during the first 4 h of incubation. The maximum of formation was obtained at 320–360 mOsmol. Secretion of membranes ended at 700–800 mOsmol. On the other hand, 1 mm of peritrophic membranes per hour was still formed at 200 mOsmol. The width of the band pattern observed in polarized light after staining of the membranes with Congo red showed maximum values at an osmolarity of the incubation medium between 320–360 mOsmol. In some cases values corresponded to those found in vivo. Below and above this osmolarity the width of the macrounits was 25–35% smaller than those formed in vivo. Above 430 mOsmol or below 280 mOsmol the width of the macrounits decreased more and more, sometimes up to 50%, and irregularities occurred. Becker et al. (1975) assumed that osmotically regulated enzymes might be present in the cells secreting peritrophic membranes. The mechanism could work by the interconversion of a bulk polymer and the respective monomer. The effects of a variation in pH were investigated at an osmolarity of 340 mOsmol and at 27 °C. In this case the osmolarity was adjusted with sorbitol. The maximum formation was at a pH of 6.8–7. The secretion of peritrophic membranes was observed down to pH 5, and up to pH 8.5. At the optimal pH of 6.8–7, the width of the anisotropic band pattern was 70% of that formed in vivo. In the range between pH 5.3–6.3 and 7.6–8.7 the width of the periodic units decreased considerably; sometimes only irregular patterns were found.

The influence of temperature on the formation rate of peritrophic membranes was investigated by adjusting the osmolarity with sorbitol to 340–360 mOsmol and by adjusting the pH to 6.8. A maximum formation rate of 3.6 ± 0.7 mm of peritrophic membranes per hour was obtained at temperatures between 26–28 °C. The width of the periodic units was reduced about 20% compared with those formed in vivo. In the temperature range 18–26 °C the units were only slightly smaller than those formed in vivo, whereas in the nonphysiological range of 33–40 °C the units were considerably smaller than in vivo, and irregularities occurred.

The addition of 20-hydroxy-ecdysone to the incubation medium resulted in a marked increase in the formation rate: 4.5–5.5 mm of peritrophic membranes were formed per hour. But the width of the periodic units was reduced about 30–50% compared with those formed in vivo. Electron microscopy revealed that the chitin-containing part of the membranes had considerably increased in width and that the formation of the electron-dense layer on the lumen side of the inner peritrophic membrane (PM 1) was disturbed. The sysnthesis of peritrophic membranes in vitro was inhibited by the addition of *Cecropia* juvenile hormone to the culture medium (Fig. 54; Becker 1978b). At concentrations of 5–10 μg juvenile hormone/ml culture medium the formation rate diminished considerably. With increasing concentration the formation rate decreased linearly. The anisotropic crossbands were up to 80% smaller than in the controls. Large parts of the membranes showed irregularities in the band pattern. The formation zone of the cardia was partly damaged. Although the dry weight of such membranes was reduced, the inner membrane was thickened

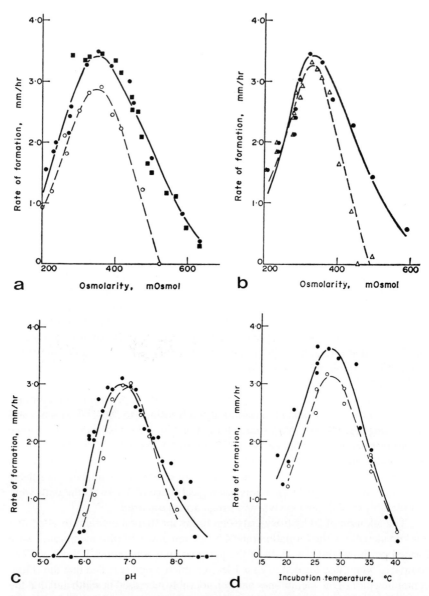

Fig. 53a-d. Formation of peritrophic membranes by isolated cardiae of *Calliphora erythrocephala* in vitro under different conditions. *a,b* The rate of formation is plotted versus increasing osmolarity of the incubation medium; pH 6.8, 27 °C. *a* • Leloup's medium adjusted with sucrose; ■ Leloup's medium adjusted with sorbitol; ° modified Tyrode solution adjusted with sucrose. *b* • Leloup's medium with glucose, osmolarity adjusted with sucrose; Δ Leloup's medium without glucose, osmolarity adjusted with sucrose. *c* The rate of formation is plotted versus the pH of the incubation medium. • Leloup's medium and ° modified Tyrode solution were adjusted to 340 mOsmol/l and held at 27 °C. *d* The rate of formation is plotted versus increasing temperature. • Leloup's medium and ° modified Tyrode solution were adjusted to 340 mOsmol/l and a pH of 6.8 (After Becker et al. 1975)

and appeared to be loosely packed in the matrix part. The electron-dense layer was disturbed considerably.

Normally, both hormones occur together with different concentrations in the hemolymph. Therefore, it might be expected that the detrimental effects of one hormone on the formation of peritrophic membranes could be prevented or at least reduced by the addition of both hormones simultaneously. However, a combination of both hormones using various concentrations of juvenile hormone and a constant concentration of 5 µg 20-hydroxy-ecdysone/ml culture medium had no compensation effect for the detrimental effects of juvenile hormone but rather an intensification.

The chemical composition of the culture medium seems to be of minor importance for the formation rate, but of great importance for the structural properties of peritrophic membranes formed in vitro. This concerns the band pattern as well as the fine structure. In vivo 6.2 mm of peritrophic membranes per hour is formed, whereas in vitro only 3.6 mm of peritrophic membranes per hour is secreted at 26–28 °C. As hunger reduced the formation rate to up to 2 mm of peritrophic membranes per hour, it seems likely that the incubation medium is still not optimal with respect to some factors. However, Becker et al. (1975) discussed as further possible mechanisms the regulation of the formation rate by an osmotically regulated enzyme system, the temperature regulation of the fly, or still other mechanisms. In a further study Becker et al. (1976) used a flow chamber for the formation of peritrophic membranes in vitro in order to avoid the accumulation of metabolic substances during prolonged incubation. This resulted in a nearly constant production of 3.5 ± 1.4 mm of peritrophic membranes per hour at 27 °C during the first 8–10 h of incubation. However, after about 10 h the production of peritrophic membranes decreased and stopped after about 35 h of incubation. In peritrophic membranes secreted during the first 8 h, the anisotropic band pattern reached about 80% or nearly the width observed in membranes formed in vivo. The fine structure of the membranes did not differ from those formed in vivo. No qualitative or quantitative disparities between peritrophic membranes grown in vitro during the first 8 h of incubation and those secreted in vivo could be observed.

Further results will be reported in the following chapter.

3.3.8 Factors Influencing the Formation of Peritrophic Membranes

Only a few investigations consider the effects which inhibit the protein or chitin synthesis or induce hormonal effects on the formation of peritrophic membranes either in vivo or in vitro. Such experiments were designed to obtain more insight into the chemical composition, the interdependence of the different components and the transport function of peritrophic membranes.

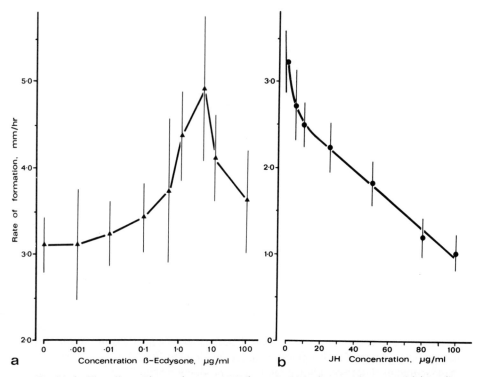

Fig. 54a,b. The effect of increasing concentrations of *a* 20-hydroxy-ecdysone and *b* juvenile hormone on the formation rate of peritrophic membranes by isolated cardiae of adult blowflies, *Calliphora erythrocephala*, in vitro. Each *point* represents the mean of 5 experiments; the *bars* represent confidence limits of 95% (After Becker 1978b)

Dimilin

First of all, an inhibitor of chitin synthesis, Dimilin or Diflubenzuron, was tested. Dimilin affects chitin synthesis indirectly by preventing the activation of chitin synthetase (Marks and Leighton 1980). In continuously operating systems, e.g. the formation zones of peritrophic membranes, chitin synthetase is in an activated form. Therefore, Dimilin can affect only small portions of newly formed enzyme. Consequently, the formation rate decreases by only about 30% after the application of Dimilin (Becker 1978b).

Clarke et al. reported already in 1977 that this chemical blocks chitin synthesis during the production of peritrophic membranes in *Locusta migratoria*. It had a twofold effect, an increase in fibrous appearance and a reduction in weight. The ratio of chitin and protein was unchanged, although the absolute amounts of each component were reduced. The amount of chitin appears to be rather high: 46.2% after treatment with Dimilin as compared with 44.4% in controls.

In adult mealworms, *Tenebrio molitor*, the peritrophic envelope was completely lacking after 8 days of treatment with Dimilin (Soltani 1984). In this case the insecticide was incorporated in unknown but probably rather high doses, as it was administered with food using 0.1, 0.5, 1 or 10 g Dimilinper kg wheat flour containing 5% dried yeast.

Becker (1978b) investigated the effects of Dimilin on the formation of peritrophic membranes of the adult blowfly, *Calliphora erythrocephala*, in vivo and in vitro under the same conditions as described in Sect. 3.3.7. In vivo the formation rate at 27 °C was about 50% higher as compared with the rate in vitro (Becker 1978b). All Dimilin concentrations resulted in the same reduction of the average formation rate. After treatment with 10 µg Dimilin/ml culture medium, the total dry weight of peritrophic membranes synthesized in vitro was 24% lower than in controls. The anisotropic bands of peritrophic membranes formed in vivo were only slightly smaller than those of controls. But in vitro the width of the crossbands was reduced by about 50%. Electron micrographs show that peritrophic membranes formed after application of Dimilin in vivo and in vitro are much thicker than in controls. The microfibrils appear to be more loosely packed than in normal membranes. The electron-dense layer on the lumen side of the inner peritrophic membrane appears to consist of a more or less large number of layers. Peritrophic membranes formed after the application of Dimilin showed an increase in permeability (Zimmermann and Peters 1987).

Polyoxin D

Polyoxin D is a competitive inhibitor of chitin synthetase due to its structural similarity to UDP-N-actylglucosamine (see Muzzarelli 1977). Becker (1980) investigated the effects of Polyoxin D on cardiae of adult blowflies, *Calliphora erythrocephala* in vitro. During the first 5 h of incubation, the formation rate of peritrophic membranes decreased linearly about 35% in the range of 5–10 µg Polyoxin D per ml culture medium. But with further increasing concentrations of Polyoxin D no further reduction in membrane length was found. If the rate of formation was expressed in µg membrane dry weight per hour, it decreased linearly with increasing concentrations of Polyoxin D in the range of 5–100 µg/ml culture medium. Becker observed a reduction of membrane dry weight and an inhibition of 88% of the chitin synthesis during in vitro tests after the application of 100 µg/ml culture medium. However, higher concentrations of Polyoxin D did not result in a significantly greater reduction of the chitin content in these peritrophic membranes. In controls the protein accounts for about 31% and the chitin for 7.5% of the dry weight of these peritrophic membranes. After incubation with a Polyoxin D-containing culture medium, the total amount of chitin decreased, but the relative proportions of chitin and protein in the dry weight remained constant. The synthesis of peritrophic membranes ended after 10 h of incubation in culture medium with 5 to 100 µg Polyoxin D per ml. The anisotropic crossband pattern was present in membranes grown in media with the addition of 1 or 10 µg Polyoxin D, but the width of the bands was reduced by about 30% compared with the controls. As these crossbands were still present after treatment with 0.5 N NaOH at 100 °C, Becker assumed that the bands might represent specially arranged chitin-containing microfibrils. Peritrophic membranes which were synthesized in media containing more than 10 µg Polyoxin D per ml did not show the anisotropic crossband pattern but only large disturbed zones. Electron microscopy showed that the three membranes formed under these conditions were thicker than in controls. Depending on the concentration of Polyoxin D, there was an increase in fibrous appearance. The electron-dense layer on the lumen side of the inner peritrophic membrane showed stacks of lamellae with irregular zones. This proteoglycan-rich but chitin-free layer is regarded as an important structure with respect to the permeability properties of these peritrophic membranes. Zimmermann and Peters (1987) investigated the effects of Polyoxin D on the permeability properties of peritrophic membranes grown in vivo. Fluorescein isothiocyanate labeled dextran fractions were fed as markers and the permeability of the peritrophic membranes in the isolated gut was observed with a fluorescence microscope as described by Peters and Wiese (1986). These authors observed an abrupt increase in the permeability of peritrophic membranes which were formed in vivo after the fly had been fed 20 µg Polyoxin D per ml glucose solution. Higher concentrations of up to 100 µg Polyoxin D per ml had only little additional effect on the permeability. In spite of the morphological alterations which were observed by Becker (1980) in membranes which had been formed in vitro in media containing 1 or 5 µg Polyoxin D per ml, no facilitation of the passage of larger FITC-labeled Dextran fractions was observed.

Captan

Captan is used as a fungicide since it is able to inhibit chitin synthesis. Marks and Sowa (1976) reported that it is also capable of inhibiting chitin synthesis in insects. Moreover, it also inhibits numerous enzymes which leads to interferences in phosphate, amino acid and carbohydrate metabolism. The effects of Captan on the formation of peritrophic membranes of the adult blowfly, *Calliphora erythrocephala*, were investigated by Becker (1978) and Zimmermann and Peters (1987). If Captan was added to the culture medium, up to 0.5 µg/ml incubation medium, it resulted in a decrease in the formation rate by about 50% compared to the control. In higher concentrations the decrease in the formation rate was slower. Captan also had a detrimental effect on the structure of these peritrophic membranes. Anisotropic crossbands did not occur in membranes formed under the influence of Captan, but instead large irregularities were observed. The membranes were thickened considerably, the electron-dense layer on the lumen side of the inner peritrophic membrane was partly absent or transformed into stacks of layers which sometimes formed loops. In the matrix part the fibrous appearance increased considerably. The effects of Captan on the permeability of these peritrophic membranes were comparable to those induced by Polyoxin D (Zimmermann and Peters 1987).

Calcoflour

Calcoflour is used as a brightener in washing agents since it is able to insert between certain carbohydrate chains, for instance cellulose and chitin. Maeda and Ishida (1967) found out that it is able to inhibit the formation of chitin-containing microfibrils as well as other glucans by integration into the macromolecules during their formation. It may also destabilize the protein moiety of peritrophic membranes. Zimmermann and Peters (1987) investigated the effects of Calcoflour on the fine structure and permeability of the peritrophic membranes of the adult blowfly, *Calliphora erythrocephala*, after feeding with this compound. Treatment with Calcoflour had no detrimental effects on the formation cells in the cardia or the appearance of the peritrophic membranes formed. The permeability of these peritrophic membranes to FITC-labeled dextran fractions was raised compared with the controls. However, it cannot be explained at present why 20 µg Calcoflour per ml feeding solution caused a more prominent increase in permeability than a concentration of 100 µg/ml.

Cycloheximide

Cycloheximide inhibits the protein synthesis of eukaryotic cells. Its effects on the fine structure and permeability of peritrophic membranes was investigated in the adult blowfly, *Calliphora erythrocephala*, by Zimmermann and Peters (1987). One to 2 h after the injection of 10 µl of a solution containing 3–5 µg cycloheximide per ml, electron microscopy revealed conspicuous changes in the fine structure of the peritrophic membranes (Fig. 55). The electron-dense layer on the lumen side of the innermost peritrophic membrane (PM 1) was enlarged and appeared as a multilayered structure with loops and branchings. The degree of hypertrophy decreased in the middle part of formation zone 1 until only one layer was visible, which became increasingly fragile and discontinuous. At the beginning of this formation zone no electron-dense layer or its precursors was observed. Therefore, the application of cycloheximide caused first an increased secretion of proteins and proteoglycans, which decreased, while the newly secreted membrane was moved towards the midgut proper. The matrix of the inner peritrophic membrane showed a more fibrillar appearance than the controls. More cellular debris was embedded in the matrix of this membrane than under normal conditions. The authors assumed that in the formation zone there probably exists a protein pool which is emptied similarly to that observed in other proteoglycan-secreting cells. Cycloheximide is believed to inhibit its replenishment. Therefore, when all proteins available in this pool are secreted, no further proteins are available for the synthesis of complete glycoproteins or proteoglycans. Thus, the electron-dense layer becomes discontinuous and secretion finally ends. As cycloheximide is an inhibitor of general biosynthesis of proteins, it might also effect chitin synthesis by inhibiting enzyme synthesis. The appearance of the second peritrophic membrane was less changed. However, the outer third membrane (PM 3) was enlarged considerably, up to tenfold in width compared to controls. Unexpectedly, no change in permeability occurred after the injection of 100 µg/ml cycloheximide or after feeding the inhibitor

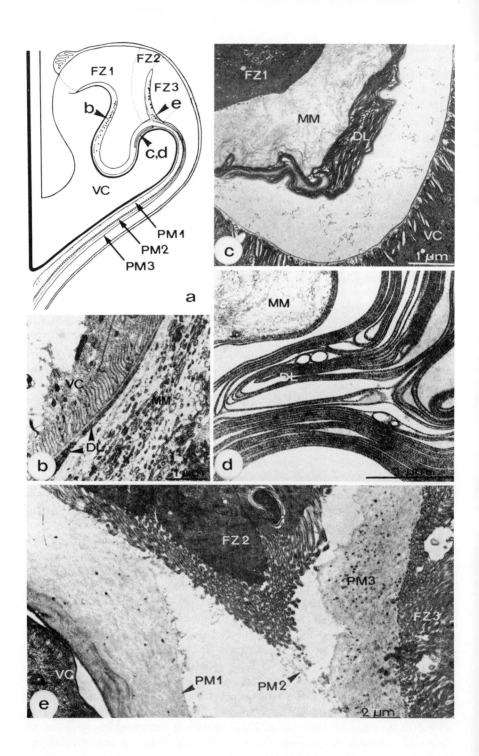

116

in concentrations of 10, 20 or even 50 μg/ml glucose solution. However, if cyclo-heximide was fed in a concentration of 100 μg/ml, an increase in permeability of the peritrophic membranes was obtained in the posterior but not in the anterior part of the midgut. Higher concentrations of the inhibitor could not be used as they proved to be toxic.

α-Amanitin

After the injection of α-amanitin, an inhibitor of RNA polymerase, peritrophic membranes were no longer formed in *Anopheles stephensi*. The proteolytic activity in the midgut was drastically reduced to 17% compared to controls (Berner et al. 1983).

5-Hydroxytryptamine (5-HT), Serotonin

Vincent (unpubl.) found that in the locust, *Locusta migratoria*, and the cockroach, *Periplaneta americana*, the formation of peritrophic membranes is enormously increased after the injection of 5-hydroxytryptamine (5-HT, serotonin). However, the cells involved in the synthesis of peritrophic membranes appeared to be ex-hausted about 10–30 min after injection. Vincent, therefore, assumed that in the midgut cells of locusts and cockroaches the components of the peritrophic membranes appear to be in readiness for secretion.

In the adult blowfly, *Calliphora erythrocephala*, a continuous secretion of peritrophic membranes occurs and not a secretion in batches as in the locust or cockroach. After the injection of 10 μl of 7.5×10^{-5} mol/1 5-HT a less pronounced effect, an increase of 25% of the formation rate, was observed (Finke and Peters 1989). The formation rate was only related to the length of membrane but not to its thickness. The resulting membranes did not differ in fine structure, protein and glycoprotein spectrum, and protein and chitin content compared with the controls. Gel electrophoresis revealed a variety of proteins and glycoproteins with apparent molecular weights in the range 12.5 and 220 kDa. Proteins with apparent molecular

---

Fig. 55a-e. The formation system of the peritrophic membranes of the adult blowfly, *Cal-liphora erythrocephala* (Diptera), is rather fragile. Similar disturbances as shown here were observed after the application of inhibitors of protein or chitin synthesis and during in vitro formation of peritrophic membranes by the isolated cardia in unsuitable incubation media. *a* Schematic representation of the right half of the cardia of *C. erythrocephala* to indicate the position of the different sections. *FZ 1–3* Formation zones 1–3; *PM 1–3* peritrophic membranes 1–3; *VC* valvula cardiaca. *b* After the application of cycloheximide considerable amounts of cellular debris are visible in the matrix material (*MM*) of the inner peritrophic membrane. *DL* Electron-dense layer. *c,d* In the lower part of formation zone 1 the electron-dense layer (*DL*) of the inner peritrophic membrane is hypertrophied considerably and the matrix material (*MM*) shows a more fibrillar appearance. *e* After the application of cycloheximide the amount of material of the second peritrophic membrane (*PM 2*) is reduced and its typical arrangement disturbed. The third peritrophic membrane (*PM 3*) is extremely thick (After Zimmermann and Peters 1987)

weights between 80 and 85, of 48 and between 12.5 and 20 kDa stained particularly intensively. Glycoproteins were analyzed after blotting gels to nitrocellulose. The blots were incubated with lectin-gold conjugates in order to evaluate the different terminal sugar residues. The specificity of binding was controlled by competition with the respective sugars. Glycoproteins with terminal N-acetylglucosamine residues had apparent molecular weights of 200, 100 and 90 kDa, respectively; those with mannose or glucose residues 100, 70, 60, 46, 36 and 20 kDa, respectively; and those with galactose or N-acetylgalactosamine residues had apparent molecular weights of 180 and 100 kDa. Terminal fucose residues were not found. The amount of readily soluble protein extracted from membranes with SDS and urea amounted to $8.3 \pm 0.8\%$ of dry weight peritrophic membranes from control flies and $9.2 \pm 2.2\%$ of dry weight of peritrophic membranes from flies treated with 5-HT. Poorly soluble proteins extracted with 1 N NaOH amounted to $33.8 \pm 5.9\%$ of dry weight in controls and $34.7 \pm 4.2\%$ in flies which had been injected with 5-HT. The total protein content of peritrophic membranes amounted to 42.7% in controls and 43.7% in injected flies. The peritrophic membranes had a chitin content of $9.7 \pm 1.6\%$ in controls and $8.2 \pm 3.3\%$ in flies injected with 5-HT. The width of the anisotropic band pattern was wider than in controls. At a concentration of 5×10^{-5} mol/l the length of membrane formed during a certain period increased by about 14%, but at a concentration of 1×10^{-6} mol/l no effect was observed. Concentrations higher than 2.5×10^{-2} mol/l proved to be toxic. The flies were unable to move, and the rate of membrane formation was reduced by about 18% compared with the controls. The thickness and fine structure of these membranes were altered. This shows that higher concentrations of 5-HT affect the mechanism of formation of peritrophic membranes in the cardia. It has to be considered that these concentrations regard the amount of injected 5-HT and that 5-HT is diluted and rapidly degraded in the hemolymph.

The mechanisms controlling the formation of peritrophic membranes are still completely unknown. 5-HT induces several hormonal and neurotransmitter effects on organs and physiological functions of insects (Miller 1975; Collins and Miller 1977). To mention just relevant functions, it has effects on the secretion of saliva secretion as well as muscle contractions in the gut and heart. 5-HT may have either a direct or an indirect effect on the formation mechanism of peritrophic membranes. It may influence the rate of secretion as well as muscle contractions of the cardia, which could mean the transport mechanism for the newly formed peritrophic membranes.

Chapter 4

Degradation of Peritrophic Membranes

There are only a few papers dealing with the problems of degradation. Many species seem to void their food residues as fecal pellets or strands (see Sect. 6.7) which may have a characteristic form and size. In some cases the fecal strand may protrude from the anus with considerable length as in the pelagic mysis larvae of the shrimp, *Solenocera atlantidis* (Sect. 6.7) or undisturbed fly larvae (Sect. 3.3.6). On the other hand, in many species a degradation of peritrophic membranes seems to occur, probably mainly in the hindgut. This has been investigated only in a few examples.

In earthworms like *Lumbricus terrestris, L. rubellus, Allolobophora chlorotica, A. caliginosa,* and *Eisenia foetida* peritrophic membranes are rarely found in the hindgut (Vierhaus 1971). Therefore, Vierhaus tried to determine the role played by the type of food ingested and the degree of degradation of peritrophic membranes. No statistically warranted differences occurred in the amount of peritrophic membranes present in the first 90 segments of worms which had been fed either cellulose tissue, cellulose powder, soil or coarse sand which had been washed several times. However, statistically warrented differences were observed in the following segments, especially the last 20 segments. The highest degree of degradation was found in worms which had been fed soil. Therefore, an influence of soil components on the efficiency of digestion might be assumed. The difference between cellulose powder, on the one hand, and soil or sand, on the other, might be explained by mechanical abrasion which should be more intense by the coarse sand and the sand-containing soil than the cellulose particles. Moreover, chemical degradation should be possible, as the chitosan test showed that there is a diminuation of chitin in the peritrophic membranes if those from the midgut were compared with those from the hindgut of the earthworm. The decrease in the amount of peritrophic membranes by about 50% in the hindgut as compared with the midgut in earthworms which had been fed cellulose is an indication that chitinase might be involved in degradation. Chitinase has been demonstrated in the gut of *Lumbricus terrestris* by Jeuniaux (1963). Experiments in vitro showed that chitinase is able to degrade peritrophic membranes from different parts of the gut even at a pH of 6.8. After 6 h the membranes were destroyed to such a degree that no chitosan test was possible. In the gut of the earthworm a pH between 6.8–7.8 has been measured. Chitinase has an optimum at about 5.1. At pH 7 only half of its activity is still effective.

Therefore, mechanical and chemical degradation seems to be responsible for the considerable reduction in the amount of peritrophic membranes which occurs in the hindgut of the earthworm.

Mechanical degradation of the peritrophic membranes of adult but not of larval Diptera is said to be done with the small teeth on the rectal pads (Engel 1924; Graham-Smith 1934; Wigglesworth 1972). This assumption is a common view in textbooks. However, in the adult fleshfly, *Sarcophaga barbata*, it could be demonstrated that mechanical fragmentation already occurs earlier in the middle of the hindgut by teeth which are present on the cuticular ridges. The secretion of peritrophic membranes starts immediately after emergence of the flies. The rear end of the initial secretion is twisted and tight for all impermeable contents of the membranous sac. This changes as soon as the secretion of membranes proceeds and this part has been voided. Subsequently, the rear end is open. In *Sarcophaga barbata* the hindgut is provided with a cuticle which has different types of teeth. Moreover, this part of the hindgut has a strong circular muscle layer. At the beginning of the hindgut a dilatation, the pylorus, is present into which urine is discharged by the Malpighian tubules. The flow of midgut contents and urine into the hindgut proper is regulated by the valvula pylorica. The following part has several longitudinal folds which are more pronounced in the posterior part and finally reduce its lumen considerably. The cuticle of this part is studded with small teeth (Fig. 56a). They consist of a squamiform basal part and end with a fine tip. The peritrophic membranes are intact until they reach this part of the gut. The teeth are able to crush the membranes into small pieces during peristaltic movements of this part of the hindgut. However, pieces of the inner peritrophic membrane still show the typical layers, an electron-dense luminal layer and the thick electron-lucent part with fine microfibrils in it. The next part of the hindgut starts with another valve, the valvula rectalis, followed by a dilatation, the rectum. In *Sarcophaga barbata* it has, as in other flies, four cone-shaped pads which are responsible for the resorption of water. Their cuticle is studded with teeth (Fig. 56b). Their broad, basal part tapers into a single tip in the basal parts of the papillae, or have up to four tips in the distal region of the papillae. Unless they have been cleaned by several washings, the rectal pads are covered with bacteria (Fig. 56c). Frayed pieces of both peritrophic membranes were observed in this part of the hindgut. However, in other specimens the hindgut may be empty or filled with intact membranes. This shows that peritrophic membranes are transported in this part of the gut either normally or under the conditions of preparation.

Chemical degradation of the pieces of peritrophic membranes is still a matter of speculation in Diptera. As soon as the peritrophic membranes are fragmented in the hindgut, proteases could gain free access to degradable parts of the peritrophic membranes. Perhaps recycling of proteins and certain carbohydrates is useful to the fly which feeds on nectar and other carbohydrate-containing food. Recycling might not be necessary for larvae since these use food rich in proteins. Chitinase activity has not yet been demonstrated in the hindgut of flies or their larvae.

→

Fig. 56a-e. Degradation of peritrophic membranes in the hindgut of the adult fleshfly, *Sarcophaga barbata. a* Scanning electron micrograph of bristles with long and fine tips which occur in the anterior part of the hindgut and seem to be responsible for the fragmentation of PMs; 7400x. *b* Scanning electron micrograph of a cleaned rectal pad with different types of bristles; 830x. *c* Scanning electron micrograph of an uncleaned rectal pad covered with bacteria; 7600x. *d* Intact PM *1* and *2* in the midgut; 53 000x. *e* The remnants of mechanically disrupted and probably chemically degraded peritrophic membranes in the hindgut; 11 400x (After Nagel and Peters 1991)

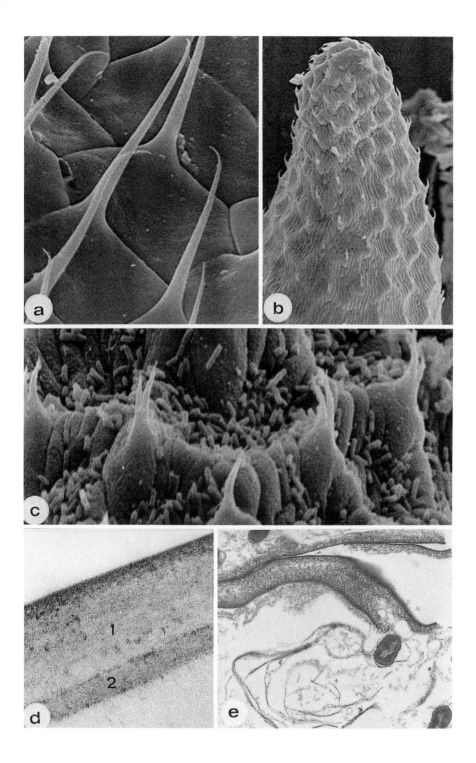

In the larvae of blowflies and fleshflies no degradation of PM seems to take place in the hindgut. If these larvae are allowed to crawl undisturbed in water, the peritrophic membrane protrudes from the anus and reaches considerable length outside the larva. For a fly a similar situation would be a hinderness during flight. In larvae the peritrophic membrane is destroyed normally by fragmentation in the surrounding food.

Chemical degradation of peritrophic membranes has been demonstrated only in larvae of Lepidoptera. Ono and Kato (1968b) investigated the chitinolytic properties of enzymes obtained from culture filtrates of a bacterium of the genus *Aeromonas*. If healthy larvae were induced to spit digestive juice, no chitinolytic activity could be found. But if larvae which suffered from flachery disease were treated in the same way, they produced only a small amount of digestive juice. Their peritrophic membranes proved to be partially dissolved, and the fragments plugged the esophagus. Experiments in vitro showed that the digestive juice of larvae suffering from flachery disease was able to dissolve intact peritrophic membranes. Several genera of bacteria were isolated from the digestive juice of these larvae. *Aeromonas* had the highest chitinolytic activity followed by *Pseudomonas, Alkaligenes, Klebsiella* and *Serratia*.

Silkworm larvae, *Bombyx mori*, which had been reared on an artificial diet, were infected with *Streptococcus faecalis* AD-4. These bacteria became attached to the peritrophic envelope and began to multiply. During 4 days after inoculation, the pH of the digestive fluid decreased, whereas the pH of the hemolymph increased. After 2 or 3 days large colonies had formed. By 6 days after infection the peritrophic membranes were dissolved, the midgut cells were damaged and the larvae died.

In order to increase the susceptibility of larvae to infection with insecticidal microorganisms, Daoust and Gunner (1979) tried to degrade the peritrophic membranes of the gypsy moth, *Lymantria dispar*, with chitinase of bacterial origin in vivo and in vitro. However, the alkalinity in the midgut is not compatible with the low pH required for the activity of chitinases. Therefore, an attempt was made to find bacteria which are able to lower the pH in the midgut of gypsy moth larvae. Bacteria isolated from the bacterial flora of the larval gut proved to be superior to various isolates from other sources. These were combined with chitinolytic bacteria for a synergistic microbial attack on gypsy moth larvae.

Chapter 5

Chemical Composition

5.1 General Remarks

Peritrophic membranes normally contain chitin, proteins, glycoproteins and proteoglycans (acid mucopolysaccharides). In addition to these, we have only some quantitative values for the hexosamine, glucose and uronic acid contents. It has been suggested that most of the glucosamine comes from the chitin moiety, whereas most of the remaining membrane weight comes from glycoproteins and proteoglycans. Richards and Richards (1971) pointed out that if one adds up these values the total never exceeds 50–70%. The question is, where is the discrepancy?

Lipids were only found by Gander (1968) from peritrophic membranes of adult *Aedes aegypti* and *Anopheles stephensi* by using thin-layer chromatography, but not after staining with Sudan black B. Becker could never find even traces of lipids in the peritrophic membranes of larval or adult blowflies, *Calliphora erythrocephala* (unpubl. results).

5.2 Chitin

5.2.1 Properties

There are three principal skeletal materials in organisms which occur as microfibrils: cellulose, chitin and collagen. In animals cellulose is very rare. It occurs only in the test of Tunicata and under pathological conditions in the connective tissue of man; in both cases it has been identified by X-ray diffraction.

The name chitin was introduced by Odier (1823). It is derived from the Greek word for a tunic. It is a poly-β-$(1 \rightarrow 4)$-N-acetyl-D-glucosamine. Chitin does not occur alone, but always as a chitin-protein complex. Chitin-containing microfibrils are embedded in a protein matrix. Collagen represents the reverse principle with protein microfibrils in a polysaccharide matrix. The chitin content seems to be of importance for the tensile strength of peritrophic membranes, whereas the proteoglycans might protect the membranes against enzymes and have an influence on the permeability properties. Information on the tensile strength of chitin is very sparse. It is in the order of cocoon silks of the silkworm, *Bombyx mori*, and the garden spider, *Araneus diadematus*, but considerably smaller than that of cellulose fibers or dragline silk of *Araneus diadematus*. The tensile strength of dry purified chitin was 9.5×10^7 N m^{-2}, but that of wet chitin was only 1.8×107 N m^{-2}; that of balkenlagen chitin of a bettle was 5.8×10^8 N m^{-2} and that of the apodeme of a spider, *Tegenaria agrestis*, was 2.7–6.8 × 10^7 N m^{-2} (compiled by Wainwright et al. 1976). It has been assumed that the differences between dry and wet material may be due to a different degree of polymerization, a preferred orientation, or both.

The comprehensive investigations of Jeuniaux (1963) have shown that chitin is probably of very ancient origin, or in the language of phylogenetic systematics, it is an *ancestral or plesiomorphic feature*. Hennig (1950, 1982) stated that conformity in plesiomorphic features (symplesiomorphy) cannot establish the basis for a phylogenetic relationship. This is only possible if there is conformity in apomorphic or derived features (synapomorphy). Therefore, the occurrence of chitin itself in different groups is a symplesiomorphy and cannot be used as an argument for a phylogenetic relationship, although it must be admitted that it appears at first sight to be tempting. Only specialized chitin-containing structures (synapomorphies) can be used for phylogenetic speculations.

The ability to synthesize chitin has probably been lost several times in the animal kingdom. In this context the Mollusca are a good example. Ancestral groups have chitin associated with proteins and calcium carbonate in their shells (Peters 1972), and they have chitin in their peritrophic membranes (Peters 1968). A somewhat controversial example is represented by the Chordata. Peters (1966, 1968) reported the occurrence of chitin in the peritrophic membranes of several species of Tunicata (Ascidiacea and Thaliacea) on the basis of the results of the chitosan test, but Rudall and Kenchington (1973) denied this due to X-ray diffraction and infrared absorption studies. However, they used preserved material and may have confused peritrophic membranes and food material. A reinvestigation is urgently required! Jeuniaux (1963) assumed that the enzyme, which is responsible for the synthesis of chitin, chitin synthetase, is absent in all Deuterostomia due to a genetic defect which may have occurred at the beginning of their evolution. If the Tunicata still have chitin in their peritrophic membranes, this cannot be true. Fungi and possibly Tunicata are the only groups which are also able to synthesize chitin and cellulose as well.

Chitin is usually regarded as poly-N-acetyl-D-glucosamine or officially as (1 → 4)-acetamido-2-deoxy-β-D-glucan. However, this simple view has to be revised (Muzzarelli 1977). Chitin from different sources may be acetylated to a different degree. This could be of importance for the water-binding capacity of chitin, since it is assumed that for every missing acetyl group there are two water molecules.

Delachambre (1969) demonstrated that the amino groups are responsible for a positive PAS reaction of chitin just after secretion, before it is linked to protein. In stabilized and sclerotized cuticle the PAS reaction is negative. If protein is removed with KOH at 100 °C over 24 h, unsclerotized cuticle becomes PAS-positive probably because reacting amino groups were unmasked.

X-ray diffraction is supposed to be the most reliable method for the detection of chitin (Rudall and Kenchington 1973).

Chitin is regarded as a long, unbranched chain of N-acetylglucosamine residues. X-ray diffraction studies have revealed three polymorphic forms which differ in the arrangement of the molecular chains within the crystal cell (Rudall 1963; Rudall and Kenchington 1973). In α-chitin the chains are antiparallel, in β-chitin all chains have the same orientation, and in γ-chitin there are groups of three chains with different

124

orientation. The three polymorphic forms of chitin have been found in different parts of the squid, *Loligo*: the beak contains α-chitin, the pen β-chitin, and the gastric shield or cuticular lining of the stomach γ-chitin. In all octopods examined, the gastric shield showed γ-chitin. Muzzarelli (1977) suggested that this indicates that the three forms of chitin are relevant to functional, not systematic aspects. Peritrophic membranes from a dragonfly, a locust, a mantis, a cockroach, a silkworm larva and a sawfly larvae had γ-chitin, whereas larvae of Hymenoptera Aculeata (honeybee, bumblebee and wasp) had α-chitin. Several species of weevils (Coleoptera) which use their peritrophic membranes to form cocoons had β-chitin in these membranes (Rudall and Kenchington 1973).

5.2.2 The Chitosan Reaction

The chitosan test was introduced by van Wisselingh (1898) for the detection of chitin in Fungi. It depends on the alkaline hydrolization of chitin chains and the reaction of the resulting oligomers (chitosan) with potassium iodide under acidic conditions (Richards 1951). However, it is not often realized that first of all the protein matrix has to be removed by alkaline treatment. It has often been asserted that the chitosan test is unreliable, but this view may be due to the fact that it had not been realized that the hydrolysis of the different chitin-protein complexes needs different times.

In connection with the most widely used test for the presence of chitin, the chitosan test, and its alleged uncertain reliability, it is of great importance that in some chitins other sugar components than glucosamine and glucose were found after acid hydrolysis. In the epicuticle of an arachnid, *Palamneus swammerdami*, Krishnan et al. (1955) found galactose. Stegemann (1963) reported that galactose and mannose were the products of hydrolysis of cuttlebone of *Sepia* (Cephlopoda), and in a species of the Bryozoa, *Scrupocellaria berthelotti*, the end products of hydrolysis were galactose and rhamnose. Perhaps these unexpected sugar components were responsible for the unsatisfactory results of Hyman (1958) with the chitosan test. Until now there is only one report on such "foreign" sugars in the chitin moiety of peritrophic membranes. Adult *Anopheles stephensi* secrete peritrophic membranes 30–72 h after a blood meal. Interestingly, these gave an atypical chitosan reaction: in *Anopheles sephensi* and *A. gambiae* the color was reddish rust-brown, whereas peritrophic membranes of the yellow fever mosquito, *Aedes aegypti*, were dark reddish-violet. Thin-layer chromatography revealed that the hydrolysates of the peritrophic membranes of *Anopheles stephensi* contained galactosamine and galactose, whereas those of *Aedes aegypti* contained glucosamine and glucose (Berner et al. 1983). It might be expected in a structure of such widespread occurrence like peritrophic membranes that more such examples exist. The chitosan test was modified several times, for instance by Campbell (1929) and Waterhouse (1953a). The latter used a small pressure container made from stainless steel to enable the hydrolysis of even small and delicate peritrophic membranes. The hydrolysis of chitin-containing material is achieved with KOH. Originally, a 60% solution was used at 160 °C or 20 min in a sealed tube. However, it works also with 40% at 60 °C. After hydrolysis the material is washed carefully and acidified with sulfuric acid. When Lugol's solution (iodine + potassium iodide + water = 1:2:300) is added, chitosan obtains a red to violet color. Chitosan consists of smaller chains than chitin and can be regarded as the deacetylation product of chitin or as a primary aldehyde. The molecular weight of chitin is in the range of 10^6 daltons and that of chitosan is in the range of 10^5 daltons (Muzzarelli 1977).

The acetyl groups of chitin are easily removed even with dilute alkali, whereas the amino groups are exceptionally stable in 50% NaOH at 160 °C (Muzzarelli 1977). In contrast to chitin, chitosan is soluble in dilute acetic and formic acid. Therefore, after iodine treatment of the chitosan test the membranes can be transferred to 3% acetic acid where the color disappears as the chitosan is dissolved; usually the peritrophic membranes do not dissolve, as a considerable portion of chitin still exists after this treatment.

If the chitosan test gives a negative result or produces a different color of the reaction product, this may have several reasons:

1. There may be no chitin in the respective material.
2. There may be only a very small amount of chitin.
3. The matrix may be dissolved and the chitin may be dispersed.
4. Other sugar components than glucosamine may be present (see above).
5. KOH is a very viscous substance which is difficult to remove by washing with water, especially in crevices and capillary tubes. If it has not been removed completely, there is no color change when the iodine solution is applied. Therefore, the material should be washed finally with several changes of sulfuric acid if the yellow color of the iodine solution fades. Prakasam and Azariah (1975) recommended a saturated solution of potassium iodide to deacetylate chitin. After washing Lugol's solution is applied in a pH range of 1.4–4; sulfuric acid is avoided in this case.
6. The time needed for an adequate degradation and deacetylation of chitin in order to give chitosan depends on the respective material (Peters 1968; Ravindranath and Ravindranath 1975).

Many failures may be due to the wrong timing of this process. Peters (1968) reported that peritrophic membranes of *Salpa democratica* and *Tethys vagina* (Tunicata) had to be treated for 1 week with 40% KOH at 60 °C until a positive chitosan reaction was obtained. The extremely long duration of hydrolysis could have depended on the resistance of the associated protein to alkaline degradation.

Despite the fact that several authors emphasized that the chitosan test was unreliable, it was of great value in the history of investigating peritrophic membranes. Until the 1930s the following question was controversial: Are the peritrophic membranes only mucus or do they really contain chitin? Today, this is only of historical interest. Wigglesworth (1930) and von Dehn (1936) were the first to use the chitosan test with great success to investigate the occurrence of chitin in the peritrophic membranes of various insects. Their results and those of others were the basis for the opinion of Waterhouse (1953a,b) that the chitin content is the only generally accepted criterion of peritrophic membranes. The results obtained with the chitosan test on peritrophic membranes have also been confirmed by other methods, e.g. X-ray diffraction and enzymatic degradation.

5.2.3 Enzymatic Degradation

One can remove the protein moiety, digest the chitin enzymatically with chitinase and then determine quantitatively the N-acetyl-glucosamine content (de Mets and Jeuniaux 1962 see Jeuniaux 1963). Peritrophic membranes are immersed in liquid nitrogen, hydrophilized and weighed with an electronic microbalance. Then proteins and glycoproteins are extracted with an aqueous solution of 5% SDS (sodium dodecyl sulfate) and 8 M urea for several hours at room temperature. After washing the membranes with distilled water, another extraction in 1 N NaOH for 2 h at 100 °C follows which removes proteoglycans. Both extracts can be used for the determination of the protein content after Lowry et al. (1951) or Bradford (1976). The extracted membranes are used for the estimation of the chitin content according to the method of Jeuniaux (1963). The purified chitin is first degraded with chitinase. The membranes are incubated for 10 h at 37 °C in a medium containing 1 mg chitinase in 2 ml 0.15 M citric acid buffer and 0.3 M disodium hydrogenphosphate buffer. After centrifugation the resulting chitobiose in the supernatant is degraded with β-glucosidase. The resulting N-acetylglucosamine is determined after the colorimetric method of Reissig et al. (1955) using p-dimethyl benzaldehyde; the extinction is determined at 585 nm.

The three methods mentioned above only allow one to demonstrate the presence of chitin, but not its localization in the tissues, i.e. it allows a gross localization only.

5.2.4 Histochemical Methods for the Localization of Chitin

A histochemical method based on the dichroitic staining of chitin with Thiazin red (Füller 1965) did not prevail. In 1969 Benjaminson proposed a localization method with chitinase labeled with a fluorescent dye, either with fluorescein isothiocynate (FITC) or with lissamine rhodamine. However, this method has two drawbacks: (1) the FITC fluorescence is in the same wavelength range as the intrinsic fluorescence of chitin. (2) No specific inhibitor of chitinase was known which could prove the specificity of the reaction.

The lectin wheat germ agglutinin (WGA) has a binding site for N-acetyl-glucosamine. It is structured in such a way that it is able to bind a sequence of three β-(1 → 4)-linked residues of this sugar (Allen et al. 1973). Therefore, it has a strong affinity to oligomers and polymers of N-acetylglucosamine as chitin. The specificity of binding can be proved by competition with the respective acetylated oligomers.

Ultrastructural localization of biologically important proteins, like lectins, glycoproteins, antibodies, etc., can be achieved by the use of colloidal gold as an electron-dense marker (Geoghegan and Ackerman 1977; Horisberger and Rosset 1977; Horisberger 1979; Goodman et al. 1981; Peters et al. 1983; Roth 1983; Polak and Varndell 1984). Colloidal gold is obtained for instance by the reduction of tetrachloro auric acid (Frens 1973). The grain size depends on the conditions of reduction and has a very small degree of variability (Horisberger 1979). The grains are negatively charged and easily loaded with the respective protein at its isoelectric point. The conjugates must be stabilized (for further details, see Horisberger 1979).

As WGA is a relatively small molecule with a molecular weight of about 40 000 daltons, it has to be coupled with bovine serum albumin (BSA) by glutaraldehyde, according to Horisberger and Rosset (1977), before it can be used for labeling the colloidal gold. The negatively charged gold particles are labeled with WGA-BSA at pH 7.4. The labeling conditions were already described in detail by Geoghegan and Ackerman (1977), Horisberger (1979), Goodman et al. (1981), Peters et al. (1983) and Peters and Latka (1986). The gold sol is stabilized with polyethylene glycol (PEG) with a molecular weight of 20 000 daltons. The labeled gold sol is stored at 4 °C. Before use, the content of colloidal gold in a probe can be checked by measuring the optical density at 520 nm (Goodman et al. 1981), in order to have, after appropriate dilution, a comparative amount of gold in the incubation medium.

A very important point is the embedding resin. Epoxy resins do not react with the water-soluble reagents which are needed. More useful are embedding media like the water-soluble Lowicryl K4M or LR White. Sections of both can be incubated with colloidal gold without prior etching.

Peritrophic membranes are isolated and cleaned from food and food residues. Then they are washed several times with phosphate-buffered saline, fixed in 0.2% glutaraldehyde buffered with 0.1 mol/l cacodylate pH 7.4 for 1–2 min, and washed carefully in phosphate-buffered saline 3×10 min. Afterwards they are fixed again in 0.5% glutaraldehyde buffered with 0.1 mol/l cacodylate for 60 min. After 3×10 min washing in cacodylate buffer the material is sampled in a small drop of 3.5% agar, and treated for 45 min with 50 mM ammonium chloride to block aldehyde groups (Roth 1983). After washing and dehydration in a graded series of ethanol and in propylene oxide, the material is embedded in either LR White or Lowicryl K4M. The embedded material can be used either for TEM or LM.

5.2.4.1. Electron Microscopy (TEM)

Ultrathin sections are put on grids which were coated with Formvar and carbon. Then they are equilibrated with phosphate-buffered saline + 1 mM $CaCl_2$ + 1 mM $MgCl_2$ for 5–10 min and incubated with WGA-BSA gold conjugate, e.g. diluted 1:50; the most appropriate degree of dilution must be tried out. After several washes with distilled water the sections are stained with uranyl acetate and lead citrate and observed with an electron microscope (Fig. 60b,c).

Specificity of binding can be controlled by competitive inhibition with 0.2–0.3 N-acetylglucosamine or with 5–20 mM triacetyl chitotriose. Allen et al. (1973) could show that competition of WGA with triacetyl chitotriose is 3000 times more efficient than with N-acetyl glucosamine. The acetyl groups are of great importance in this respect. Higher oligosaccharides did not show an increase in effectiveness. Allen et al. assumed that the structure of the binding site of WGA is such that it is able to bind a sequence of three N-acetylglucosamine residues. Therefore, it has a strong affinity to oligo- and polymers of N-acetylglucosamine like chitin. However, binding may be almost impossible if chitin is masked by cuticle proteins.

For competition WGA-BSA gold and the respective sugar are mixed for 10 min; then the sections are covered with a large drop of this medium. N-acetylglucosamine is only able to inhibit binding of WGA to glycoproteins, but not to chitin; the latter is possible with triacetyl chitotriose (Peters and Latka 1986). Subsequently, another test should be added, incubation of sections with chitinase in citric acid buffer pH 5.4 prior to incubation with WGA-BSA gold conjugate.

5.2.4.2 Light Microscopy

Semithin sections are transferred to siliconized slides. The slides are equilibated in phosphate-buffered saline (PBS) + 1 mmol/l $CaCl_2$ + 1 mmol/l $MgCl_2$ and then covered with a large drop of WGA-BSA gold solution diluted about 1:5 with PBS. It is unnecessary to remove LR White or Lowicryl K4M. Moreover, 0.1% BSA (fatty acid-free) should be added to the WGA solution in order to avoid unspecific binding. After 60 min of incubation at room temperature in a humidity chamber, the slides are washed three times for 10 min with distilled water. Latensification of the small gold grains is achieved with a physical developer Intense-silver enhancer (Janssen, Beerse, Belgium) for about 4–6 min which can be used in normal light. After rinsing in distilled water, the results can be checked microscopically in order to determine the optimum time of incubation. Binding sites of the silver-intensified gold marker appear more or less dark gray (Fig. 60a). Specificity of binding is controlled as described previously.

Springall et al. (1984) recommended counterstaining with hemalum. However, the dark blue stain hides the more or less dark gray appearing parts with silver-latensified gold. Azocarmine, which is part of Heidenhain's Azan method, appears to be more suitable. It stains nuclei red and tinges also other parts of cells if it is used as a progressive stain. Azocarmine G (0.1 g) is boiled shortly in 100 ml of distilled water. Then the solution is filtered and 1 ml of acetic acid is added. Twenty to 60 min are needed for staining; more rapid staining is possible at 60 °C. The stained sections are rinsed with distilled water, dried, cleared in xylene and mounted in a suitable resin.

The results obtained with these methods will be described in other sections.

5.3 Proteins and Glycoproteins

The matrix of peritrophic membranes contains chitin microfibrils embedded in proteins, glycoproteins and proteoglycans. Quantitative analyses are hampered by the possibility of contamination of the membranes with food, food remains and enzymes.

5.3.1 Amino Acids

The first amino acid analysis was done by Ono and Kato (1968). They isolated and cleaned peritrophic membranes from 30 000 silkworm larvae, *Bombyx mori*, amounting to 1.8 mg. There are five or more layers of peritrophic membranes in the midgut of these larvae. They consist of patches of membrane which are secreted by delamination from the whole midgut epithelium. A total protein content of 34.3% by weight and remarkably high percentages of basic amino acids were found: 21.66% arginine and 8.13% lysine; glycine with 3.88% was also abundant (Fig. 57).

More recently, the amino acid composition of the larvae of another two species of Lepidoptera, *Manduca sexta* and *Orgyia pseudotsugata*, was investigated by Adang and Spence (1982). The results are given in Table 10. The total protein content was the same, 35% by dry weight. The major amino acid of peritrophic membranes of *Manduca sexta* was glycine (8.8% by weight), followed by aspartate (4.9%), glutamate (3.7%) and proline (3.1%). In the peritrophic membranes of *Orgyia pseudotsugata* the threonine content was highest (5.1% by weight), followed again by glutamate (4.5%), aspartate (4.2%) and proline (3.2%).

Although the protein content of the peritrophic membranes of these two species was the same, there were considerable differences in the protein spectra (Table 11). In *Manduca sexta* only a single protein with 62 000 daltons contained carbohydrate, whereas in *Orgyia pseudotsugata* two proteins with 220 000 and 140 000 daltons stained with PAS. The content of hexosamines of the membranes of *Orgyia pseudotsugata* was approximately twice that of *Manduca sexta* (Table 10).

Zimmermann et al. (1975) investigated the tube-like peritrophic membranes of adult blowflies, *Calliphora erythrocephala*, which are secreted continuously at a

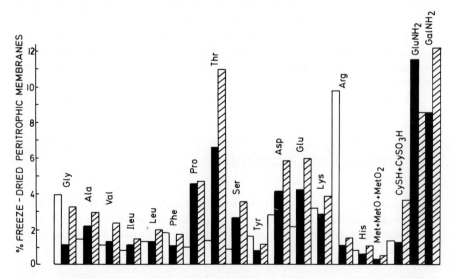

Fig. 57. Diagrammatic representation of the amino acid and amino sugar composition of peritrophic membranes of the silkworm, larva of *Bombyx mori* (Lepidoptera) (*white columns*; after Ono and Kato 1968), PM 1 and 2 of adult blowflies *Calliphora erythrocephala* (*black columns*; after Zimmermann et al. 1973), and the peritrophic membrane of larval C. *erythrocephala* (*hatched columns*) (Zimmermann and Klein, unpubl. results). (After Peters 1976)

Table 9. Quantitative determinations of chitin and protein (% percent by weight: L, larva, I, imago, adult)

Species	Chitin (%)	Protein (%)	Author
Dixippus morosus	3.7	47	de Mets and Jeuniaux (1962)
Aeschna cyanea/L	10.6	21	de Mets and Jeuniaux (1962)
Bombyx mori/L	12.9	42	de Mets and Jeuniaux (1962)
Bombyx mori/L	–	34.3	Ono and Kato (1968a)
Calliphora erythrocephala I grown in vivo	7.5	36	Zimmermann et al. (1975)
Calliphora erythrocephala I grown in vitro	7.5	30.6	Becker (1980)
Manduca sexta/L	–	35	Adang and Spence (1982)
Orgyia pseudotsugata/L	–	35.2	Adang and Spence (1982)

Table 10. Comparison of the amino acid and amino sugar composition of the peritrophic membranes of the larvae of three species of Lepidotera: *Bombyx mori* (Ono and Kato 1968a), *Manduca sexta* and *Orgyia pseudotsugata* (Adang and Spence 1982), and Diptera: adult *Calliphora erythrocephala* (Zimmermann et al. 1975) and larvae of this species (Zimmermann and Klein, unpubl. results). Determination of hexosamines for *Bombyx* by de Mets and Jeuniaux (1962). All values in % dry weight

	Bombyx larvae	*Manduca* larvae	*Orgyia* larvae	*Calliphora* larvae	*Calliphora* adults
Aspartic acid	2.79	4.9	4.2	5.8	4.1
Threonine	1.27	1.1	5.1	11.0	6.6
Serine	0.78	1.5	1.8	3.5	2.6
Glutamic acid	2.05	3.7	4.6	5.9	4.1
Proline	0.93	3.1	3.2	4.6	4.5
Glycine	3.88	8.8	2.0	3.2	1.1
Alanine	1.38	1.6	2.5	2.9	2.1
Valine	1.06	1.3	1.8	2.3	1.2
Isoleucine	0.77	1.1	1.1	1.4	1.0
Leucine	1.26	1.0	1.7	1.9	1.2
Tyrosine	1.64	0.9	1.2	1.1	0.7
Phenylalanine	1.76	0.9	0.9	1.7	1.0
Half cystine	1.28	2.5	2.3	3.6	1.2
Cysteic acid	–	–	–	–	–
Arginine	9.66	0.9	1.1	1.4	1.0
Lysine	3.07	0.6	0.6	3.8	2.8
Histidine	0.74	0.9	0.8	1.0	0.5
Methionine				0.4	0.2
Protein	34.32	35.0	35.2	55.5	35.9
Hexosamines	14.5	9.4	17.9	20.7	20.0
Glucosamine		7.8	11.8	8.5	11.5
Galactosamine		0.7	4.8	12.2	8.5

Table 11. Protein patterns of peritrophic membranes of larvae (L) of *Manduca sexta* and *Orgyia pseudotsugata* (compiled from data of Adang and Spence 1982) and larvae (L) and adults (I) of *Calliphora erythrocephala* and *Sarcophaga barbata* (Stamm et al. 1978)

		Calliphora		Sarcophaga	
Manduca / L	*Orgyia* / L	L	I	L	I
210	220 +	200 +	200 +		200 +
190					
165					
140	140 +		130 +		
84					
76		70 +			74
66				67	
62 +		60		56	59
		50		49 +	53
37	37			42 +	
32					
	25,	25		28	
	23		21		21
	20	18		17	19,17
(14)	(14)	15	14	15	
			10		12

Molecular weights are given as apparent M in kDa. + PAS positive. In both species of Lepidoptera peptides of 14 kDa occurred at the gel front.

high rate by specialized cells of the cardia at the entrance of the midgut. They can be cleaned much more easily than the patches of membranes of the silkworm. For a single analysis 100–200 membranes were needed. There are three morphologically different membranes of different thickness (Fig. 62b,c). The third membrane which faces the midgut epithelium is a very delicate tube with a thickness between 20–50 nm. As it cannot be seen during preparation, it cannot be excluded that it may be lost during isolation and cleaning processes.

In the peritrophic membranes of adult *Calliphora erythrocephala* the protein content amounted to 36% by weight. Threonine (6.6% by weight) proved to be the predominant amino acid, followed by proline (4.5%), aspartate and glutamate (4.1%, respectively) and lysine (2.8%) (Fig. 57).

The larvae of *Calliphora erythrocephala* secrete a single, tube-like, continuous peritrophic membrane which is covered on the lumen side by bacteria, *Proteus vulgaris* and *P. morganii*, if the larvae are reared under normal conditions (Fig. 66; Peters et al. 1983). It is impossible to clean the membranes completely as the bacteria stick firmly to them. However, it is possible to obtain clean membranes by rearing the larvae under sterile conditions. Such membranes were investigated by Zimmermann and Klein (unpubl. results; Fig. 57). These peritrophic membranes had a higher protein content than those of the Lepidoptera and the adult blowfly: 55.5% by weight. As in the peritrophic membranes of the larvae of *Orgyia pseudotsugata* and the adult blowfly, threonine was the major component. It amounted to 11% by weight as compared with 6.6% in adult blowflies. Similarly, higher contents of the following amino acids were determined: glutamate 5.9%, aspartate 5.8%, proline 4.6% and lysine 3.8%.

Threonine is the major amino acid in the peritrophic membranes of three of the species which were investigated until now (Fig. 57). It is followed by high contents of the dicarboxylic amino acids aspartate and glutamate. Although it is obviously too early to draw conclusions, it is tempting to speculate on the possibility that these amino acids might be used for linking proteins with carbohydrate to form glycoproteins or to link proteins with the chitin moiety. Only five monosaccharides and five amino acids are known to be involved in such linkages. Threonine and serine can bind to acetylgalactosamine, asparagine to acetylglucosamine, hydroxylysine to galactose, serine to xylose and hydroxyproline to arabinose.

There are large quantitative differences between the amino acid composition of peritrophic membranes and that of a range of other structural proteins such as wool, elastin, collagen, silks, etc. (Hunt 1970).

5.3.2 Proteins

Stamm et al. (1978) were the first to investigate the protein and glycoprotein spectrum of peritrophic membranes (Fig. 58). Larvae and adults of the blowfly, *Calliphora erythrocephala*, and the fleshfly, *Sarcophaga barbata*, were reared under sterile conditions to obtain clean peritrophic membranes without bacterial contaminations.

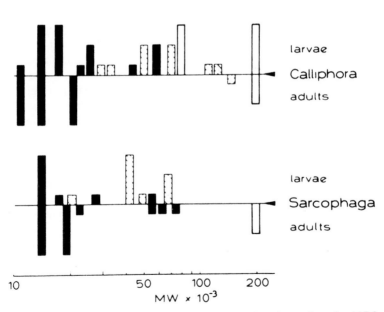

Fig. 58. Schematic representation of the protein band pattern after electrophoresis of SDS extracts from peritrophic membranes of larval and adult blowflies, *Calliphora erythrocephala*, and fleshflies, *Sarcophaga barbata*. *Open columns* PAS-positive only; *black columns* Coomassie blue-positive only; *hatched columns* sensitive to both. The height of the column represents the relative staining intensity (After Stamm et al. 1978)

Electropherograms of food and food residues in the larval midgut showed that impurities adhering possibly to the peritrophic membranes after isolation and cleaning processes at least did not interfere with the major protein and glycoprotein bands of the peritrophic membranes. Interferences with minor bands could not be excluded completely; therefore, the minor bands were not taken into account in Fig. 58. Electron microscopic controls proved that only the innermost peritrophic membrane (PM 1) of adult *Calliphora erythrocephala* was investigated. SDS was used for extraction as it is known to release proteins and glycoproteins from plasma membranes without breaking covalent bonds. It is well known that glycoproteins, due to their different carbohydrate content, show anomalous behavior in SDS-gel electrophoresis. Therefore, only apparent molecular weights are available for glyco-proteins by this method.

Only the bulk of proteins with a molecular weight of 18 kDa could be found in similar amounts in both stages of both species. The protein patterns of the larval stages of a given species were more uniform than the glycoprotein patterns. The main bands had molecular weights between 10 and 30 kDa. Some minor bands were found in the range of 30 to 80 kDa.

5.3.3 Glycoproteins

A carbohydrate fraction with an apparent molecular weight of 200 kDa was present in the peritrophic membranes of both stages of *Calliphora* and larvae of *Sarcophaga*, but not of adult *Sarcophaga* (Fig. 58). It could be detected only by PAS staining. The larval peritrophic membranes of both species had some glycoprotein bands with positive staining properties for PAS and Coomassie blue in the range between 30 and 80 kDa. There were no glycoprotein bands present in this range in the adult stages. As sialic acids proved to be a functionally important and characteristic component in many glycoproteins of vertebrates and in some invertebrate phyla, an approach was made to detect it in peritrophic membranes by the resorcinol method of Heuser et al. (1974). However, the results were negative in normal peritrophic membranes as well as in SDS extracts.

A comparison of the glycoprotein patterns of peritrophic membranes revealed great differences between the adult and the larval stage within the two species (Fig. 58). However, some similarities exist between the larval and the adult stages. Both species have similar living habits. The variability of the protein and glyco-protein patterns seems to indicate that there may be special requirements in each stage and species (Stamm et al. 1978).

Terminal sugar residues of glycoproteins as well as sugar-binding proteins can be detected with suitably labeled colloidal gold in sections used for electron (Figs. 59–63) and light microscopy (Fig. 62a; see the preceding Sect. 5.2) as well as in blots of electropherograms (Fig. 64; Towbin et al. 1979; Burnette 1981; Rohringer and Holden 1985). The sensitivity of the method is greatly enhanced by using silver latensification with Intense (Janssen, Beerse, Belgium) (Moeremans et al. 1984). This is needed because the amount of material is very small. The specificity is controlled by competition with the respective sugars and sugar deriva-tives.

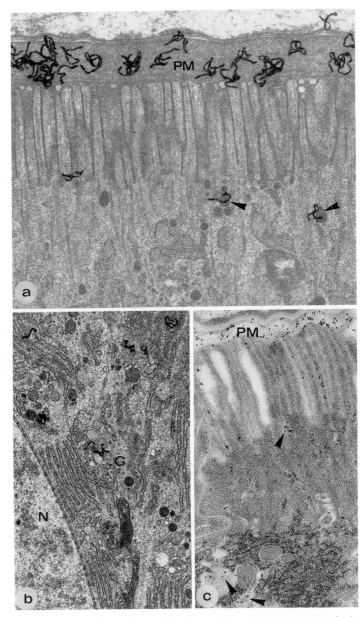

Fig. 59a-c. Comparison of the results of radioactive and lectin-gold labeling of the single peritrophic membrane of the larva of a blowfly, *Calliphora erythrocephala. PM* Peritrophic membrane. *a* After the injection of ^{3}H-galactose into larvae, peritrophic membrane and vesicles with rather electron-dense contents (*arrowheads*) are labeled in electron microscopic autoradiographs; 21 800x. *b* In the vicinity of Golgi fields similar vesicles are labeled and probably transported to the surface of the cell. *G* Golgi field; *N* nucleus; 23 600x. *c* The lectin soy bean agglutinin (SBA) can be used to localize N-acetylgalactosamine residues more precisely than in autoradiographs. If sections of the cardia are incubated with SBA gold conjugate, vesicles with rather electron-dense contents (*arrowheads*), the space between the microvilli and the newly formed peritrophic membrane are labeled; 25 100x

The pattern of terminal sugar residues on the surface of peritrophic membranes could be shown by incubation of isolated peritrophic membranes with different lectin-gold conjugates. The results of comparative investigations on several genera of Nematocera (Dörner and Peters 1988), the tsetse fly, *Glossina morsitans* (Fig. 60), on larvae and adults of the fleshfly, *Sarcophaga barbata* (Fig. 61), and on larvae (Fig. 59) and adults (Fig. 62) of the blowflies, *Calliphora erythrocephala* and *Protophormia terrae-novae* (Fig. 60), are compiled in Tables 12 and 13.

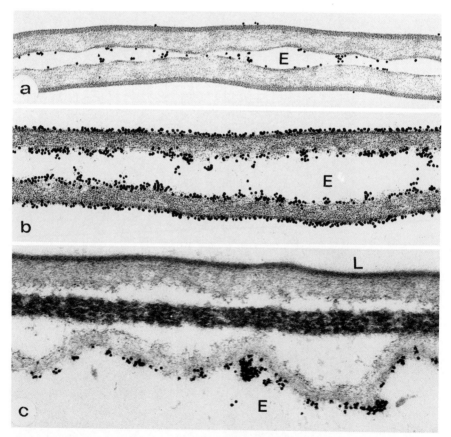

Fig. 60a-c. Electron microscopic localization of sugar residues with some lectins labeled with colloidal gold shows quite different results. *E* Epithelium side; *L* lumen side. *a* The isolated single peritrophic membrane of the tsetse fly, *Glossina morsitans*, is labeled with soy bean agglutinin (SBA)-gold conjugate almost exclusively and specifically on the epithelium side. SBA demonstrates the presence of N-acetylgalactosamine residues; 34 000x. *b* If the isolated peritrophic membrane of the tsetse fly is incubated with wheat germ agglutinin (WGA)-gold conjugate, both sides of the membrane are labeled specifically, which means that N-acetylglucosamine residues are present on both sides; 41 000x. *c* The same procedure gives quite different results with the three morphologically different peritrophic membranes of a blowfly, *Protophormia terraenovae*. Only the third membrane is labeled on the epithelium side (*E*); 41 000x

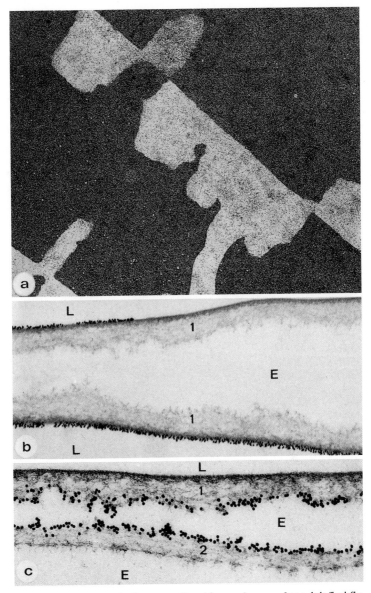

Fig. 61a-c. A peculiar situation occurs in the two peritrophic membranes of an adult fleshfly, *Sarcophaga barbata. a* After incubation of isolated inner peritrophic membranes (PM *1*) with the lectin peanut agglutinin (PNA) labeled with colloidal gold, the membrane tube was cut longitudinally and the membrane was spread on a grid gold to obtain a surface view. Labeling proved to be interrupted for more or less wide areas; 3900x. *b* In isolated inner peritrophic membranes (PM *1*) which were incubated with soy bean agglutinin (SBA)-gold, the label is found exclusively and specifically on the lumen side (*L*) and shows interruptions corresponding to those in *a*; 28 500x. *c* If isolated peritrophic membranes (PM *1* and *2*) were incubated with WGA-gold and sectioned, the label occurred in PM *1* on the epithelium side (*E*) and in PM *2* on the lumen side; 22 800x

Table 12. Lectins and terminal sugar residues of glycoproteins in the peritrophic membranes of larvae and adult Diptera as revealed by electron microscopic investigations using labeled colloidal gold as an electron-dense marker

Larvae of Nematocera[b]

Terminal sugar[a]	Detected with lectin	Peritrophic membranes of							
		Aedes		*Anopheles*		Culex		*Odagmia*	
		L	E	L	E	L	E	L	E
GlcNAc	WGA	+	+	+/–	+	+	+	+	+
GalNAc	SBA	+	+	+	+	+/–	+	+	+
GalNAc	DBA	+	+/–	+	+	–	–	+	+
GalNAc	PNA	+/–	+/–	+	+	–	–	–	–
Fuc	UEA	+/–	–	+	–	–	–	–	+/–
Man/Glc	ConA	+	+	–	–	+	+	–	+

Adults of *Glossina morsitans*

Terminal sugar	Detected with lectin	PM	
		L	E
GlcNAc	WGA	+	+
GalNAc	SBA	+/–	+/–
GalNAc	DBA	+	+
Gal	PNA	+/–	+/–
Fuc	UEA	+/–	+/–
Man/Glc	Con A	+	+

Adults of *Sarcophaga barbata*

Terminal sugar	Detected with lectin	Inner PM		Outer PM	
		L	E	L	E
GlcNAc	WGA	–	+	+	+/–
GalNAc	SBA	–	+	–	–
GalNAc	DBA	+	–	+/–	+/–
Gal	PNA	+	+/–	–	–

Adults of *Protophormia terraenovae*

Terminal sugar	Detected with lectin	Inner PM		Median PM		Outer PM	
		L	E	L	E	L	E
GlcNAc	WGA	+/–	–	–	–	–	+
GalNAc	SBA	–	+	+/–	–	–	+
GalNAc	DBA	–	+/–	+	+	–	–
Gal	PNA	–	–	+	+	–	+/–

138

Table 12. (*Continued*)

Adults of *Calliphora erythrocephala*

Terminal sugar	Detected with lectin	Inner L	PM E	Median PM L	PM E	Outer L	PM E
GlcNAc	WGA	–	–	+	+	+	–
GalNAc	SBA	+/–	+	–	–	–	–
GalNAc	DBA	–	–	–	–	?	?
Gal	PNA	+	+	–	–	–	–
Fuc	UEA	+/–	–	+	+	–	–

Larvae of *Calliphora erythrocephala*

Terminal sugar	Detected with lectin	PM L	E
Man-lectin	Peroxidase	+	–
Gal-lectin	Asialofetuin	–	–
Gal	PNA	–	–
GalNAc	SBA	+	–
GlcNac	WGA	–	+
Fuc	UEA	–	–

[a]Man-lectin, mannose-specific lectin; Man, mannose; gal, Galactose; GalNac, N-acetylgalactosamine; Glc, glucose; GlcNAc, N-acetylglucosamine; Fuc, L-fucose; PM, peritrophic membrane.
[b]+, Present; +/–, single gold particles bound; –, no binding; L, lumen side; E, eptihelium side.

Table 13. Incubation of sterile peritrophic membranes and midgut epithelium of *Calliphora erythrocephala* larvae with horseradish peroxidase-gold and asialofetuin gold[a]

	L	E	Mg
Asialofetuin	–	–	–
Horseradish peroxidase	+	–	–
Methyl-Man → horseradish peroxidase	–	–	–
Gal → horseradish peroxidase	+	–	–
Con A	+	–	
Methyl-Man + Glc → Con A	–	–	

[a]L, Lumen side; E, epithelium side; Mg, midgut epitheliu.

The electron-dense layers in the peritrophic membrane of larvae of *Aedes aegypti* reacted intensely with WGA-BSA gold conjugate (Fig. 63). However, this does not show that they contain chitin, as Richards (1975) and Richards and Richards (1971) maintained on the basis of extraction and staining experiments. Chitin microfibrils are not present in these layers but occur in the electron-lucent parts of the membrane. They seem to be masked heavily by proteins and therefore do not react with WGA-BSA gold. Histochemical evidence shows that the electron-dense layers contain proteoglycans (Peters 1979). In all species of Nematocera mentioned in Table 12 N-acetyl-glucosamine and N-acetylgalactosamine residues are present on the luminal as well as the epithelial side of the membranes. WGA-BSA gold is bound very intensely at the epithelial side (Fig. 63). Mannose- and galactose-specific lectins could not be found on these peritrophic membranes.

The peritrophic membrane of the larvae of a blackfly, *Odagmia ornata*, contains proteins/glycoproteins with apparent molecular weights of 140, 130, 75, 63, 54, 22, 18, 16 and 12 kDa. Those with 22, 18, 16 and 12 kDa showed a specific binding of SBA on the blots which binds to N-acetylgalactosamine residues (Dörner and Peters 1988; Fig. 64). In adult *Calliphora erythrocephala* fucose residues could be demonstrated electron microscopically on both sides of the second peritrophic membrane with Ulex agglutinin (UEA). Binding could be inhibited completely with 0.3 mol/l L-fucose. The lumen side of the innermost peritrophic membrane was only faintly but specifically labeled. As the pili of bacteria adhering to the lumen side of this membrane were labeled, too, it might be that the binding was due to the remnants of pili on the membrane and was not characteristic of the membrane itself. The third peritrophic membrane did not bind UEA.

Wheat germ agglutinin (WGA) was not at all bound to the innermost peritrophic membrane. It was heavily bound to both sides of the second peritrophic membrane, and less but specifically to the lumen side of the third peritrophic membrane. Therefore, N-acetylglucosamine residues of glycoproteins can be postulated on both sides of the second as well as on the lumen side of the third peritrophic membrane of the adult blowfly. Chitin does not seem to be accessible during incubation of intact membranes; its presence can only be demonstrated in sections (Peters and Latka 1986).

5.3.4 Lectins

Peters et al. (1983) reported for the first time evidence for the presence of a mannose-specific lectin and a corresponding glycoprotein with terminal mannose residues on the lumen side of the peritrophic membrane of the larvae of the blowfly *Calliphora erythrocephala*. The localization of this lectin was possible by labeling colloidal gold with horseradish peroxidase. The latter is a glycoprotein with terminal mannose residues which interact with mannose-specific lectins (Stahl et al. 1978). Specificity of binding was controlled by competitive inhibition by preincubation of membranes with 0.3 M methyl-D-mannoside or horseradish peroxidase (Fig. 65). After treatment of isolated and cleaned peritrophic membranes with colloidal gold labeled with horseradish peroxidase, the positive reaction could be observed already with the naked eye due to the pink color of the membranes. Transmission electron microscopy revealed a rather dense decoration of only the lumen side, but not of the epithelium side of the peritrophic membrane with the electron-dense labeled gold granules (Fig. 65). This could be shown for membranes of larvae which had been reared under sterile conditions. In larvae which were reared under normal conditions the lumen side of the peritrophic membrane was studded with two species of bacteria, *Proteus vulgaris* and *P. morganii* (Fig. 71b). They seem to adhere to the

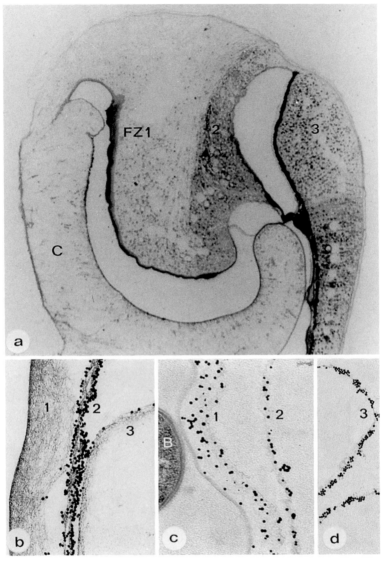

Fig. 62a. Light and *b-d* electron microscopic localization of sugar residues with the lectin WGA labeled with colloidal gold in the peritrophic membranes of adult blowflies, *Calliphora erythrocephala* (Diptera). *a* Unstained semithin section of one-half of the cardia incubated with WGA gold and afterwards silver latensification. The formation zones and peritrophic membranes are heavily labeled. *C* Cardia; *FZ 1–3* formation zones 1–3; 340x. *b* Isolated peritrophic membranes were incubated with WGA-gold, embedded in Araldite and sectioned. Both sides of the second peritrophic membrane (PM *2*) are heavily labeled, PM *3* shows only scanty labeling on the luminal side, whereas the innermost PM *1* is not labeled; 37100 x. *c,d* If sections of peritrophic membranes embedded in Lowicryl K4M are incubated with WGA gold, all three membranes are labeled except in the electron-dense layer on the lumen side of PM *1*. Competition experiments show that in this case chitin has been labeled. On the *left side* of *c* a bacterium (*B*) has been sectioned. *c* 34000 x; *d* 29000 x (After Peters and Latka 1986)

141

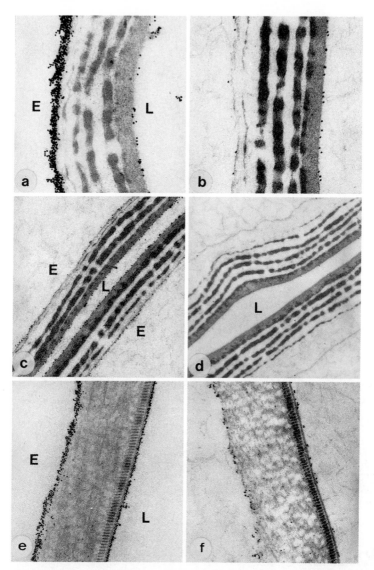

Fig. 63a-d. The peritrophic membrane of the larva of the yellow fever mosquito, *Aedes aegypti* (Diptera, Culicidae), after incubation with lectin-gold conjugates (after Dörner and Peters 1988). *a* After incubation with WGA gold the epithelium side (*E*) is heavily loaded with labeled gold particles, whereas the lumen side (*L*) shows only scanty labeling. *b* After competition with 0.3 M N-acetylglucosamine there is complete inhibition on the epithelium side but no inhibition on the lumen side. *c* PNA binds equally on both sides of the peritrophic membrane. *d* It is inhibited by competition with 0.3 M galactose. *e,f* The peritrophic membrane of a blackfly larva, *Odagmia ornata* (Diptera, Simuliidae), after incubation with lectin-gold conjugates. *e* After incubation with WGA gold. *f* Competition of WGA with 0.3 M N-acetylglucosamine results in an incomplete inhibition on the epithelium side (*E*) and scanty labeling on the lumen side (*L*)

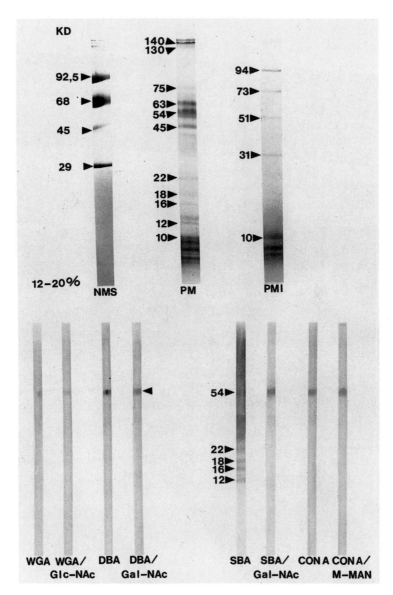

Fig. 64. The results of gel electrophoresis and blotting of SDS extracts from peritrophic membranes of blackfly larvae, *Odagmia ornata* (Diptera, Simuliidae). The apparent molecular weight in kDa is indicated. *NMS* low molecular standard; *PM* peritrophic membranes; *PMI* food residues in the peritrophic membranes. The *lower row* shows blots after incubation with lectin-gold conjugates followed by silver latensification, and the results of competition experiments with 0.2 M solutions of the respective sugars. The 54 kDa band was able to bind unspecifically all lectins tested. SBA was bound specifically by glycoproteins with an apparent molecular weight of 22, 18, 16 and 12 kDa (After Dörner and Peters 1988)

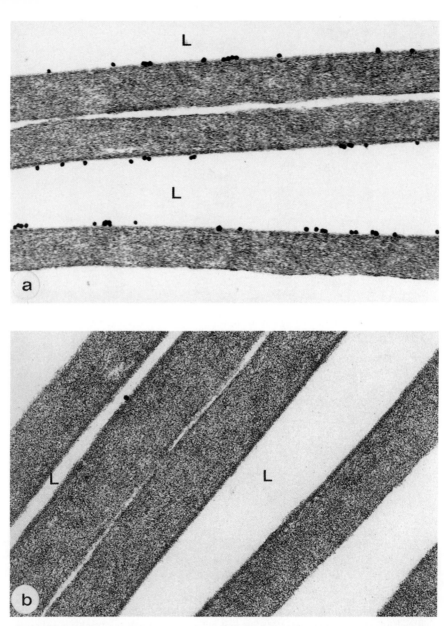

Fig. 65a,b. Evidence for the presence of a mannose-specific sugar receptor (lectin) on the lumen side of the peritrophic membrane of the larva of the blowfly, *Calliphora erythro-cephala. a* Larvae were reared under sterile conditions. Isolated membranes were incubated with peroxidase-gold conjugate. The gold marker is found nearly exclusively on the electron-dense layer of the lumen side (*L*) of the membrane. *b* Competition with 0.3 M methyl-D-mannoside abolishes nearly completely the binding of peroxidase-gold; 64 000x (After Peters et al. 1983)

peritrophic membrane by means of their fine and long pili (Fig. 66). The adherence is so strong that it is impossible to clean these membranes completely from bacteria. Membranes which had been cleaned from food and food residues were incubated with colloidal gold labeled with horseradish peroxidase. After incubation the pili of the bacteria were covered with gold (Fig. 66). Competitive inhibition with methyl-D-mannoside did not result in a complete inhibition on the surface of the bacteria; about 20% label still adhered to the pili as compared with horseradish peroxidase-gold conjugate without an inhibitor. Obviously, the pili of these bacteria also have mannose-specific lectins. After incubation of peritrophic membranes with colloidal gold labeled with the lectin Concanavalin A, which has receptors for mannose and glucose, again only the lumen side of the membrane and the pili of the bacteria were covered with the electron-dense marker. Competitive inhibition with methyl-D-mannoside and glucose resulted in a nearly complete inhibition of binding in the peritrophic membranes, but only 60–70% inhibition in the case of the bacteria. Therefore, it seems that mannose-specific lectins as well as mannose residues of glycoproteins on the lumen side as well as on the pili of these bacteria may be responsible for their mutual adherence. This could be confirmed by experiments with bacteria which were grown in culture dishes and bacteria-free peritrophic membranes from larvae which had been reared under sterile conditions. If these were brought together, only the lumen side of the membranes was covered with bacteria. After competitive inhibition with methyl-D-mannoside, the amount of bacteria adhering to the membranes was reduced considerably.

Therefore, obviously the same mechanism of mutual adherence of bacteria is present in these membranes as in *Escherichia coli* and mammalian mucosa cells (Swanson et al. 1975; Beachy and Offek 1976; Hu et al. 1976; Jones and Freter 1976; Offek et al. 1977; Aronson et al. 1979).

It is very interesting that the surface of the midgut cells has no lectins with mannose affinity. This is in agreement with the hypothesis that peritrophic membranes might be a special form of the surface coat of midgut cells which is secreted by these cells, reinforced originally and in most species by chitin-containing microfibrils, and delaminated frequently from the midgut cells in order to envelop food and food remains. In blowfly larvae a single tube-like peritrophic membrane is formed continuously in the cardia by specialized cells at the entrance of the midgut. The remaining midgut cells have no contact with the original food. Only smaller molecules and degradation products, as far as they are able to penetrate the peritrophic membranes, can reach them. Therefore, it seems possible that the midgut cells could have lost their lectins during the course of evolution in favor of the peritrophic membrane.

In adult *Calliphora erythrocephala* there is no mannose-specific lectin present on the lumen side of the innermost of the three peritrophic membranes as well as elsewhere on these membranes. On both sides of the second peritrophic membrane there are binding sites only for part of the peroxidase molecule, as competitive inhibition occurred only with peroxidase but not with either 0.3 M methyl-D-mannoside or 0.3 M galactose.

Asialofetuin is a serum protein treated with neuraminidase in order to remove terminal sialic acid residues and to expose the ensuing galactose residues. It did not react with any of the three peritrophic membranes or the single larval peritrophic

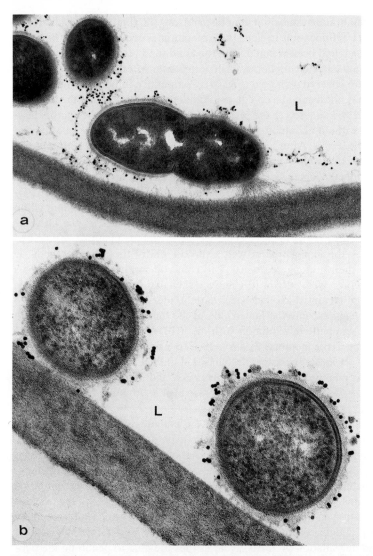

Fig. 66a,b. Evidence for the presence of a mannose-specific sugar receptor or lectin on the lumen side (*L*) of the peritrophic membrane of the larva of the blowfly, *Calliphora erythrocephala* (after Peters et al. 1983). *a* Peritrophic membranes of larvae reared under normal conditions are covered with bacteria, *Proteus vulgaris* and *P. morganii*, on the lumen side. If such membranes are incubated with gold marker labeled with horseradish peroxidase, the marker accumulates on the pili of these bacteria; 25200 x. *b* Bacteria incubated with the peritrophic membranes from larvae reared under sterile conditions bind to these membranes; 57600 x

membrane of this species. This shows that a galactose receptor seems to be absent on these membranes.

5.4 Proteoglycans

Proteoglycans (acid mucopolysaccharides) are linear macromolecules of high molecular weight with sequenced disaccharide units. Each unit consists of hexosamine and uronic acid and may be estersulfated or N-sulfated. The carbohydrate chains are covalently linked to a protein core, normally by an alkali-labile O-glycosidic bond. In glycoproteins the protein has one or many short carbohydrate side chains which are often branched. These oligosaccharides never show repeating units. They contain neutral sugars and amino sugars without sulfate groups. Uronic acids are uncommon. Only sialoglycoproteins are characterized by an acid component, sialic or N-acetyl-neuraminic acids.

Several authors described the presence of proteoglycans in peritrophic membranes of animals from different phyla (see Peters 1968) using histochemical reactions for light microscopy. Proteoglycans show a metachromatic reaction with Toluidine blue; this is due probably to a polymerization of dye molecules induced by the proteoglycans. Alcian blue was introduced by Steedman (1950) for the localization of proteoglycans. The mechanism of the reaction was investigated by Scott et al. (1964). These authors found that the addition of electrolytes is of great importance for a differentiation between several proteoglycans. For peritrophic membranes staining with the following solution for 2 h proved to be useful: seven parts of 2% Alcian blue in 3% acetic acid + one part 2 M NaCl (Peters 1968).

Scott and Dorling (1969) pointed out that proteoglycans are oxidized rather slowly by periodic acid. Therefore, they recommended a double oxidation technique. After the usual short oxidation for 1 h, sodium borohydride is used to block the aldehyde groups which are either already present or which are formed during oxidation. During a further prolonged oxidation for 24 h, aldehyde groups are formed in the proteoglycans which can be detected with the PAS method.

For the electron microscopic localization of proteoglycans several methods are recommended. Due to the negative reaction of the acidic groups of these compounds, an electron-dense reaction product is formed with cationic substances such as ruthenium red or colloidal iron hydroxide. The precipitation of proteoglycans by cetylpyridinium chloride was used by Shea (1971) for the localization with lanthanum nitrate. Pollock et al. (1970) recommended acriflavin-PTA (Geyer 1973). Peters (1979) coupled the double oxidation method of Scott and Dorling (1969) with the periodic acid thiocarbohydrazide-silver proteinate reaction of Thiéry (1967). Peritrophic membranes seem to have a uniform distribution of proteoglycans, except the highly specialized membranes of Diptera (Peters 1979). The continuous tube-like membranes of larvae and the innermost membranes of the adults of flies have an electron-dense layer on the lumen side which contains proteoglycans. It has been suggested that proteoglycans and especially this layer may be of great importance

Table 14. Hexosamine, sugar and uronic acid contents of the peritrophic membranes of three species of insects[a] (After De Mets and Jeuniaux 1962)

Species	Hexosamines	Neutral sugars	Uronic acids
Bombyx mori / larvae	14.5	0.8	1.6
Dixippus morosus	3.8	8.0	3.4
Aeschna cyanea / larvae	7.0	2.9	2.2

[a]Hexosamines in % glucosamine per weight; neutral sugars in % glucose by weight; uronic acids in % glucuronic acid per weight.

with respect to permeability (Peters 1976, 1979; Peters and Wiese 1986) and protection against enzymes (Rajulu 1971).

The first biochemical analysis was done by deMets and Jeuniaux (1962). The results are compiled in Table 14. The amount of uronic acids in peritrophic membranes of *Bombyx mori* larvae is approximately one-tenth of the content of about 15% by weight reported by Nisizawa et al. (1963). These authors dissected 30 000 silkworm larvae to obtain 18 g of peritrophic membranes. They determined the carbon dioxide produced during hydrolysis of the peritrophic membranes with 6 N HCl, assuming that all the carbon dioxide derived from glucuronic acid residues. Nisizawa et al. suggested that hyaluronic acid is the proteoglycan in these membranes as the total sulfur amounted to 3.4% by weight. The ester sulfate was only 0.17% by weight. The polysaccharide proved to be composed of equimolar amounts of glucuronic acid and glucosamine residues, as it could be degraded by bovine testicular hyaluronidase, and since the infrared spectra of the substance and cockscomb hyaluronic acid was in agreement. A reinvestigation as well as further information are urgent.

Proteoglycans could not be extracted from peritrophic membranes of larvae and adults of *Calliphora erythrocephala* and *Sarcophaga barbata* with SDS and urea for electrophoresis. This could be demonstrated by Stamm et al. (1978) by staining gels with Alcian blue or Toluidine blue, and by electron microscopy of extracted membranes. But uronic acid-containing material was extracted by 1 N NaOH over 24 h at room temperature. The presence of a high quantity of uronic acids and sensitivity to alkali is strong evidence for the occurrence of proteoglycans in these peritrophic membranes.

Chapter 6

Functions of Peritrophic Membranes

6.1 General Remarks

It can be expected that a structure which occurs as widely as the peritrophic membranes has several functions. The question is only which might be the primary and which might be additional functions. If the hypothesis is correct that peritrophic membranes are derived from the surface coat of the apical plasma membrane of midgut cells, it should have originally had properties of the surface coat, e.g. specific binding sites, like lectins (see Sect. 6.2) and glycoproteins, it might have been loaded with enzymes (see Sect. 6.3) and function as a permeability barrier (see Sect. 6.4). Reinforcement with chitin-containing microfibrils could have resulted in its use as protection against mechanical damage of the midgut epithelium by coarse particles of food and food residues (Schneider 1887) or by swelling of gut contents due to water uptake (Zhuzhikov 1964). Schneider's opinion that peritrophic membranes are used to protect the midgut epithelium from abrasive food particles is mentioned in every textbook dealing with peritrophic membranes. It is obviously overestimated and mostly believed to be the only function. However, it cannot be denied that it is a valuable function. But the conclusion is wrong that peritrophic membranes are absent in fluid feeders because in these there is no need for such protection. Waterhouse (1953a) could show that many Lepidoptera and Diptera which feed on fluids nevertheless form peritrophic membranes (Table 4). In hematophagous insects hemoglobin-like crystals with sharp edges occur in the gut after a blood meal (Berner et al. 1983; Fig. 37, 38). These might also be harmful to an unprotected midgut epithelium. Besides these possibly primary functions there are other functions which appear to be secondary: Peritrophic membranes may be responsible for a compartmentalization of the midgut (see Sect. 6.5), they may act as a barrier for parasites (see Sect. 6.6), they may be used to wrap the food residues (fecal pellets, see Sect. 6.7), or they may be used to form a cocoon for undisturbed pupation (see Sect. 6.8). A truly secondary function was observed in DDT-resistant strains of mosquito larvae, *Aedes aegypti*, (Abedi and Brown 1961). A DDT-resistant strain from Trinidad excreted nine times as much peritrophic membrane and six times as much DDT as a susceptible strain. The membrane protruding from the anus reached a length of 3 mm, about half as long as the mature larvae. However, in the absence of DDT neither the resistant nor the susceptible strain excreted any peritrophic membrane. This is an unusual type of DDT resistance due to the enhanced formation of peritrophic membrane induced by the uptake of DDT. Other DDT-resistant strains investigated, for instance from Key West and Cucuta, excreted only small amounts of peritrophic membrane. Their mechanism was not investigated in this case. In Onychophora peritrophic membranes are used not only to wrap the remains of food but also crystals of uric acid as excretion products (Manton and Heatley 1937). Obviously, the whole midgut epithelium is able to secrete peritrophic membranes; they are voided each day. Uric acid crystals appear on the inner surface of the peritrophic envelope or between the peritrophic membranes, just before they are fully raised from the midgut epithelium. The crystallization of uric acid seems to take place rapidly, probably in 1–3 h. It has not been observed to start earlier than 3 h after evacuation of the previous peritrophic envelope, and is complete in 6 h after this event. Up to 0.2 mg of uric acid is crystallized daily in the midgut of a fully grown *Peripatopsis moseleyi* or *P. sedgwicki*. As soon as the peritrophic envelope is formed, uric acid crystallization is said to be stopped. The envelope is pressed to form a narrow tube, which appears white and opaque with the contained mass of uric acid crystals. It is voided about 18 h later. Therefore, the regular formation of peritrophic membranes seems to be correlated with the mechanism of excretion in this group of animals. In late embryos which are about to free themselves from the egg membrane, the first peritrophic membranes are voided with the contained uric acid crystals, and remain in the oviduct until the birth of the young.

The constant renewal of peritrophic membranes and the constant waste of valuable material appear to be highly uneconomic, although one should bear in mind that nature may be sometimes, but is by no means always economic. There might be other reasons for this. Enzymes which are adsorbed to peritrophic membranes might be worn out after some time. Furthermore, the peritrophic membranes might be clogged by molecules which cannot penetrate or are adsorbed to them. Thirdly, lectins which may be present on the peritrophic membranes somehow could be blocked with time. In each case a constant renewal of the peritrophic membranes could counteract these effects.

6.2 Lectins and the Association of Bacteria

The gut is exposed to many foreign and sometimes harmful molecules, viruses, bacteria, protozoa, etc. Therefore, it can be expected that the plasma membrane of the midgut cells and perhaps also the peritrophic membranes bear receptors for the identification of such components in the gut. Such receptors could be carbohydrate, protein or lipid-binding membrane components.

For the first time a sugar receptor (lectin) was localized by electron microscopy in an invertebrate by Peters et al. (1983). The lumen side of the single peritrophic membrane of larvae of the blowfly, *Calliphora erythrocephala*, is densely covered with bacteria, *Proteus vulgaris* and *P. morganii*, which adhere to the membrane with considerable strength (Fig. 71b). It is practically impossible to clean the membrane completely from these bacteria. However, it is possible to obtain clean peritrophic membranes by rearing the larvae under sterile conditions. Such membranes were incubated with colloidal gold (see Sect. 3.2) labeled with either horseradish peroxidase or asialofetuin in order to find D-mannose- or D-galactose-binding lectins. Horseradish peroxidase is a glycoprotein with terminal mannose residues which interacts with D-mannose-specific lectins, whereas asialofetuin has terminal

D-galactose residues which can be used to demonstrate D-galactose-binding lectins. The specificity can be shown by competitive inhibition with the respective sugars. The lumen side of the peritrophic membrane as well as the pili of the bacteria proved to have D-mannose-binding lectins and probably D-mannose residues. Therefore, it can be assumed that there is mutual adherence between both partners. A similar system has been demonstrated already by Offek et al. (1977) to be responsible for the mutual adherence of the bacterium *Escherichia coli* and human mucosal cells. In the case of the peritrophic membrane of the blowfly larvae and the *Proteus* bacteria, the adherence is so strong that after several attempts to clean the membranes even parts of the surface of the bacteria or only pili remain attached to the peritrophic membrane (Fig. 66). If clean peritrophic membranes from blowfly larvae reared under sterile conditions were incubated with bacteria from laboratory cultures, the bacteria adhered only to the lumen side of the membranes. This could be inhibited by competitive inhibition with a α-methyl-D-mannoside.

Three points are remarkable. The lectin is distributed asymmetrically on the peritrophic membrane of the blowfly larvae; it appears only on the lumen side. The bacteria might be stage-specific symbionts as they do not appear in the midgut of adults; concomitantly there is no such lectin on either of the three peritrophic membranes of the adult flies. But the most interesting finding seems to be that the surface of the midgut cells has no lectins with mannose or galactose affinity. This favors the hypothesis that peritrophic membranes may be a special form of the surface coat of midgut cells which is secreted by these cells, reinforced in most cases by chitin-containing microfibrils, and delaminated frequently from the midgut cells in order to envelop food and food residues. In blowfly larvae the ability to secrete a single and continuous tube-like peritrophic membrane is restricted to annular groups of midgut cells in the cardia. The remaining midgut cells therefore have no contact with the original food and the food residues. Thus, it seems possible that the midgut cells could have lost their lectins during the course of evolution in favor of the peritrophic membrane.

In many other cases an intense attacment of microorganisms may be effected by similar mechanisms. For instance in the case of the larvae of the fleshfly, *Sarcophaga barbata* (Nagel and Peters, unpubl. results). An annular pad of bacteria is present on the valvula cardiaca at the beginning of the midgut (Fig. 67a,b). It is bounded by the cuticle of the valvula on one side and the PM on the other. The bacteria are rod-shaped and many of them are met in different stages of division. They are embedded in a clearly defined electron-lucent matrix (Fig. 67c). Fine filaments near the cuticle of the valvula appear to be degradation products of bacteria (Fig. 67). The species name is still unknown. Only a few bacteria were observed in the esophagus and in the anterior part of the midgut. They do not adhere to the cuticle of the valvula or the PM. Carbohydrates proved to be present on the surface of the bacteria according to the results of Thiéry's modification of the PAS method. However, none of the following lectins bound specifically to the surface of the bacteria: CBA, UEA, PNA and Con A; only WGA bound specifically. Therefore, at least N-acetyl glucosamine residues are present, but it cannot be concluded that there are N-acetyl glucosamine-specific lectins. Competition experiments proved that gold conjugate caught in the filamentous matrix was unspecific labeling. There were no mannose- or galactose-specific lectins present in the matrix or on the surface of bacteria.

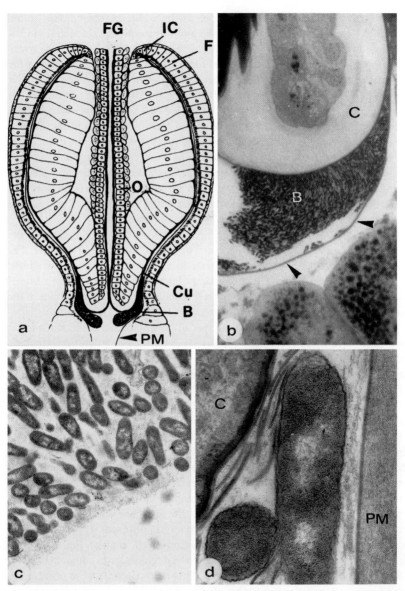

Fig. 67a-d. Bacteria which may have symbiotic functions occur regularly as an annular pad on the valvula cardiaca at the beginning of the midgut of fleshfly larvae, *Sarcophaga barbata. a* Schematic representation of a sagittal section through the cardia. *B* Group of bacteria; *Cu* cuticle of the valvula cardiaca; *F* formation zone of peritrophic membranes; *FG* foregut; *IC* imaginal cells; *O* esophagus; *PM* single peritrophic membrane. *b* The group of bacteria is bounded by the cuticle of the valvula on one side and the PM on the other side. It remains in the same position, whereas the peritrophic membrane is constantly transported to the midgut. *B* Bacteria; *C* valvula cardiaca; *arrowheads* point to the peritrophic membrane; 86x. *c* The rod-shaped bacteria are embedded in a matrix containing fine filaments; 5600x. *d* Larger filaments appear to be degradation products of bacteria. *C* Valvula cardiaca; *PM* peritrophic membrane; 28 800x

In adult *S. barbata* the cardia contains rod-shaped bacteria which adhere to the cuticle of the valvula cardiaca. Many of them are in stages of division.

In the cardia of third instar larvae of a blowfly, *Protophormia terraenovae*, only a few bacteria may be found which are not attached to the cuticle or peritrophic membrane.

In the larva of *Stomoxys calcitrans* bacteria are lacking in the foregut, but different types of bacteria occur in the food residues of the midgut lumen. However, no bacteria appear to be attached to the peritrophic membrane. The bacteria in the midgut lumen were met in division. Moreover, many empty cells occurred.

Actinomycetes cover the epithelial face of the peritrophic membrane and colonize the ectoperitrophic space of a soil-feeding termite, *Cubitermes severus* (Bignell et al. 1980). These actinomycetes are believed to be symbionts or commensals. They form filaments of different length, ranging from 0.15 to 0.8 µm, which are infrequently branched. Very fine, pili-like threads arise from the surface of these filaments and connect them with the peritrophic membrane, whereas those in the ectoperitrophic space make contact with the microvilli of the midgut epithelium. There is no information on the invasion, the way in which the actinomycetes reach the ectoperitrophic space or on the adhesive mechanisms.

Other examples where microorganisms are associated with peritrophic membranes and the cuticular lining of the hindgut are locusts, *Schistocerca gregaria* (Hunt and Charnley 1981), cockroaches, *Periplaneta americana* (Bracke et al. 1979), crickets, *Acheta domestica* (Ulrich et al. 1981) and others. However, nothing is known about the mechanisms of adhesion.

The marine wood borer, *Limnoria tripunctata* (Crustacea, Isopoda), is unique in its ability to feed on creosote-treated wooden structures. The gut of animals which were reared on untreated wood appears to be free of microorganisms. But only in animals feeding on creosote-preserved wooden pilings was a resident microflora observed on the midgut epithelium which was separated by the peritrophic membrane from other microorganisms that were ingested during wood boring (Zachary and Colwell 1979; Zachary et al. 1983). It was suggested that the bacteria circumvent the peritrophic membrane in order to reach the ectoperitrophic space. They occur also on the exoskeleton of animals feeding on creosote-treated wooden pilings, but their number decreases as soon as they are transferred to untreated wood in the laboratory. Therefore, this temporary microflora is believed to be involved in the breakdown of creosote-derived hydrocarbons.

6.3 Immobilized Enzymes

The cells of the midgut epithelium are able to secrete enzymes and peritrophic membranes originally. Only some groups, mainly of Insecta, reduced these abilities by specializing groups of cells for the secretion of peritrophic membranes (see Sect. 3.3.3). Light microscopic observations suggested that originally the two types of secretion products might be shed alternately in the anterior part of the midgut of harvestmen (Opilionida) (Peters 1967b). However, if midgut epithelium was fixed at certain intervals after the uptake of food, it could be shown by electron microscopic investigations that in *Phalangium opilio* the secretion of peritrophic membranes is induced by the uptake of food; it takes place in the anterior midgut mainly in a period between 2 and 4 h after feeding (Becker and Peters 1985).

Light and electron microscopic investigations show that the peritrophic membranes appear to be contaminated with more or less material which might contain enzymes. The enzymes could either be incorporated into the peritrophic membranes during formation, or they could be adsorbed subsequently on the fully formed membranes.

In the peritrophic envelope soluble proteases and aminopeptidases could be adsorbed to the numerous peritrophic membranes and act like immobilized enzymes. These enzymes might be lost at a lower rate than those in the lumen. Unfortunately, only aminopeptidases can be localized histochemically, but not the soluble enzymes which are present in the midgut lumen (see also Sect. 6.5). Perhaps at least some of these might also be adsorbed to the peritrophic membranes.

Petersen could show already in 1912 that the peritrophic envelope of the honeybee contains proteases. Adam (1965) found a considerable concentration of leucine aminopeptidase in the thick peritrophic envelope of *Myxine glutinosa* (Agnatha).

The presence of aminopeptidases on the peritrophic membranes of several insects was investigated histochemically by Walker et al. (1980) and Peters and Kalnins (1985). In larvae of *Drosophila melanogaster* an isozyme of leucine aminopeptidase was localized light microscopically by Walker et al. (1980) in the midgut cells, midgut lumen and peritrophic membrane. The single tube-like peritrophic membrane as well as the contents of the food adjacent to it stained darkly. Furthermore, if the entire midgut was dipped into the histochemical stain, the peritrophic membrane became a scarlet color suggesting that a very active aminopeptidase is present. Therefore, the leucine aminopeptidase is not only a membrane-bound intracellular enzyme but it may also be secreted. An increase in aminopeptidase activity seems to be induced by dietary protein; in other words, it depends upon a secretagogue stimulus given by the quantity of a specific nutrient. The authors assume the following steps of digestion which are facilitated by the compartmentalization due to the presence of a peritrophic membrane. After ingestion large protein molecules are unable to penetrate the peritrophic membrane. They are hydrolyzed in the midgut by endopeptidases like trypsin. The resulting dipeptides are able to pass the peritrophic membrane and could stimulate the secretion of aminopeptidases. The latter could associate with the peritrophic membrane. The products of protein digestion, amino acids and small peptides could penetrate the peritrophic membrane and be absorbed by the midgut cells. In the hindgut no reaction of the cells occurred, but the peritrophic membrane and the contents of the food adjacent to it gave a strong reaction. An active leucine aminopeptidase could also be demonstrated biochemically in the feces.

Peters and Kalnins (1985) could show light and electron microscopically in *Locusta migratoria* (Saltatoria), *Acheta domestica* (Saltatoria), *Forficula auricularia* (Dermaptera), *Periplaneta americana* (Blattodea), *Blaberus giganteus* (Blattodea), *Pachnoda* sp. (Coleoptera), *Apis mellifica* (Hymenoptera) and *Protophormia terrae-novae* (Diptera) that aminopeptidases occur sometimes in considerable

154

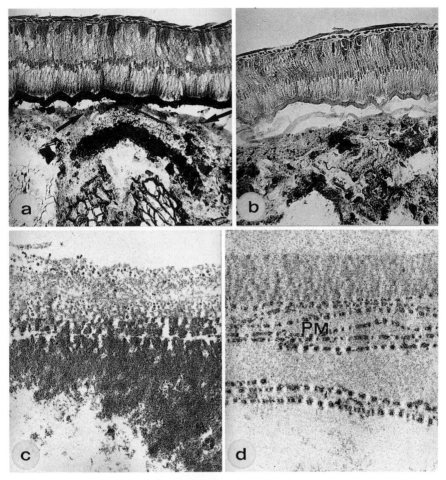

Fig. 68a-d. Evidence for the presence of aminopeptidases on peritrophic membranes (*arrows, PM*) as well as the apical parts of the midgut cells of the cockroach *Blaberus giganteus* (Insecta, Blattodea). *a* Light microscopic localization in frozen sections; 100x. *b* Control section; 100x. *c* Electron microscopy shows that aminopeptidases are adsorbed on the surface of peritrophic membranes; 12 000x. *d* Control section; 12 000x (After Peters and Kalnins 1985)

amounts on peritrophic membranes and peritrophic envelopes as well as the apical parts of midgut cells (Fig. 68). They adhere to the peritrophic membranes rather firmly even if these were washed several times and if they remained in phosphate buffer for up to 2 h. Thus, these immobilized enzymes are probably conserved for longer periods than the soluble enzymes present in the lumen of the peritrophic envelope which are shed more quickly with the feces.

In Hemiptera and Homoptera a peculiar delamination of plasma membrane material of the microvilli of the midgut epithelium occurs. Most authors assume that this is a modified form of peritrophic membranes. Marshall and Cheung called it a "plexiform surface coat", whereas others called it "extracellular membrane layers (ECML)". Cytochemical investigations proved that these structures contain at least glycophospholipids, proteins, carbohydrates and hydrolytic enzymes. Therefore, they may act as a carrier for immobilized enzymes.

In the blood-feeding bug *Rhodnius prolixus* (Heteroptera) an aminopeptidase was localized electron microscopically during blood digestion by Billingsley and Downe (1983). After a blood meal the crop of this insect is concerned initially with fluid regulation, diuresis and the activity of glycosidases which digest the cell walls of symbiotic bacteria. But it does not have a function in the digestion of proteins. Aminopeptidase occurred only as a membrane-bound enzyme of the microvilli and the complex extracellular membrane layers or plexiform peritrophic membranes of the anterior and posterior parts of the midgut. Twelve to 24 h after a blood meal a second apical membrane developed at the apical parts of the microvilli and prolif-erated subsequently to form the so-called plexiform peritrophic membrane. At the same time aminopeptidase activity was evident on the microvillar surface which increased to a maximum between 1–4 days after blood feeding. From 10–15 days after feeding the aminopeptidase activity declined gradually. The intramembrane spaces of the plexiform peritrophic membranes are believed to be comparable to the ectoperitrophic space of other insects and to be the site where terminal digestion of blood proteins occurs.

In blood-fed females of the yellow fever mosquito, *Aedes aegypti*, Graf and Briegel (1982) could show biochemically that over 50% of the total aminopeptidase activity is found in the ectoperitrophic space, 40% within the midgut cells and the remaining 10% in the blood bolus (Fig. 37b). Again, between midgut epithelium and peritrophic envelope a compartment is established in which enzymes are conserved for optimal digestion.

Eguchi and Iwamoto (1976) and Eguchi et al. (1982) investigated with biochemical methods the activity of proteases in the midgut of the silkworm, *Bombyx mori*. Proteases from midgut epithelium and peritrophic membranes proved to be immunologically identical. The concentration of the proteases, which were either adsorbed or trapped between the peritrophic membranes, was rather high compared to the concentration in the lumen of the peritrophic envelope (Fig. 69). The specific activity (tyrosine μmol min^{-1} mg^{-1} protein) amounted to 0.015 in the midgut epithelium, nearly 0.6 in the peritrophic membranes, and 0.9 in the digestive fluid. The authors suggested that the proteases are preserved or condensed in the peritrophic membranes in the course of transport from the epithelium to the lumen of the midgut, and that they are utilized effectively for the digestion of food proteins. Alkaline phosphatase from the midgut epithelium of the silkworm appeared to be loosely associated with membranes and could not be detected in the peritrophic membranes.

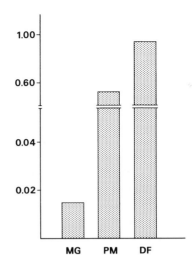

Fig. 69. Comparison of protease activity as measured by the rate of casein hydrolysis (tyrosine µmol min^{-1} mg protein^{-1}) in midgut tissue (*MG*), peritrophic membranes (*PM*), and digestive fluid (*DF*) of the silkworm, *Bombyx mori* (After Eguchi et al. 1982)

6.4 Permeability

6.4.1 Technical Problems

It has always been assumed that peritrophic membranes act as an ultrafilter since Wigglesworth (1929) wrote his famous paper on the digestion in the tsetse fly (von Dehn 1933; Schildmacher 1950; Peters 1976; Richards and Richards 1977; Bignell 1981). However, studies on their permeability properties were hindered by several difficulties. Formerly, it was impossible to obtain substances of reliable molecular weight for testing. Usually dyes were used. But these tend to form aggregations of molecules if the concentration exceeds 0.2 mol/l, which means even at concentrations which are needed for detection.

Radioactive substances have been used only occasionally (Bowen et al. 1951). Colloidal gold was formerly only obtained with different particle sizes (Schildmacher 1950; Zhuzhikov 1964). Meanwhile, it is possible to obtain particle sizes in very narrow ranges by using certain conditions during reduction of the tetrachloroauric acid solution (see Sect. 5.2). But the gold sol is not stable. Even if the particles are labeled with proteins, they are still very sensitive to the ion content of the solution and tend to aggregate or precipitate.

Markers should not be toxic. Therefore, fluorescent dyes are useful markers, since it is possible to work with extremely low concentrations, much lower than with conventional dyes. The following have been used so far: fluorescein isothiocyanate (FITC) (389.4 daltons), Procion yellow (697 daltons), Evans blue (960.8 daltons). Markers should not be decomposed by enzymes or other factors which are present in the gut lumen. Therefore, proteins are usually not useful markers, as they are easily partly degraded or completely by proteases. However, despite these handicaps ferritin (Fig. 71), hemoglobin and cytochrome c have been used. Horseradish peroxidase (M 40 kDa) is also very often used in electron microscopy because it can be detected with the diamino benzidine reaction introduced by Graham and Karnovsky (1966).

Dextrans are considered to be nearly ideal tracers as they are inert, uncharged heterodisperse polysaccharides of bacterial origin which do not seem to be decomposed in the gut lumen (Schroeder et al. 1976). They are available in a broad spectrum of fractions with different molecular weights, defined after repeated molecular sieving. Dextrans do not bind to albumin or other plasma proteins of mammals. They can be labeled with a fluorescent dye, for instance fluorescein isothiocyanate (FITC) (DeBelder and Granath 1973). Free FITC can be eliminated by repeated precipitation and dialysis. The degree of substitution is expressed as the number of fluorescein groups per 100 glucose units. Schroeder et al. (1976) found that the degree of substitution is in the range of 1% for dextrans with a mean molecular weight of 39.8 kDa and 0.5% for dextrans with M_n of 93.3 kDa. FITC is linked to the dextran by a thiocarbamoyl linkage (DeBelder and Granath 1973). This linkage proved to be stable in biological systems for about 24 h. In stored fractions the thiocarbamoyl linkage was stable at pH 4 and 20–30 °C for at least 1 month; little hydrolysis occurred at pH 9 and 20 °C after that time (DeBelder and Granath 1973). The presence of divalent cations did not alter these results. FITC-labeled dextrans can be determined by conventional fluorometry in the concentration range of 0.05–30µg/ml (Schroeder et al. 1976). Labeling of dextrans with FITC does not seem to alter the properties of the dextran molecule. Dextrans can be fed in sugar solutions and, when labeled with FITC, their way through the intestine can be followed in vivo with great precision under the fluorescence microscope (Fig. 71). The presence of injuries in peritrophic membranes due to inappropriate handling or preparation can be controlled by feeding Dextran blue which has an extremely high molecular weight of 2000 kDa. Dextran blue can be detected easily due to its intense blue color. It can be mixed with other markers if it does not interfere with them. Dextrans are readily soluble in water where the long chains form random coils of round or oval shape. Their Einstein-Stokes radii are determined by viscosity measurements. Dextran fractions are obtained by repeated molecular sieving. Therefore, a fraction does not have a definite molecular weight. The molecular weight mentioned indicates that the respective fraction contains this molecular weight as the smallest, but has a whole spectrum of larger molecules.

If even larger markers are needed, latex beads of defined sizes can be used (Fig. 70).

The inaccessibility of the ectoperitrophic space to physiological methods is a great obstacle for all investigations concerning the permeability and digestion of nutritive substances. It is impossible to collect aliquots from the ectoperitrophic space to control reliably that the markers have not changed or degraded during penetration of the peritrophic membrane and to measure quantitatively the rate of diffusion. Even with laser-light scattering the molecular size of dextran fractions from insect midgut or after they have permeated the peritrophic membrane cannot be determined due to physical reasons (Steiner, pers. comm.). Therefore, there are only arguments in favor of the stability of the labeled dextrans: (1) If degradation of dextrans occurred there could be no consistent results or differences between the species (see Table 15). (2) After the midgut epithelium had been stripped off the peritrophic membrane of an animal that had been fed with a dextran fraction which seemed to be too large to penetrate the membrane, the dextran did not penetrate even under these conditions during 1 h (Fig. 71c). If either degradation of dextrans

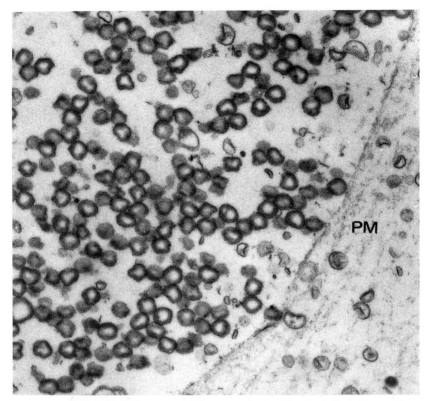

Fig. 70. Latex beads with a diameter of 327 nm are retained by the peritrophic envelope (*PM*) of the water flea, *Daphnia pulex*; 16 000x (After Hansen, unpubl. results)

or hydrolysis of the fluorescein isothiocyanate linkage had taken place in the meantime, the smaller molecules should have leaked out and this should have been observable by fluorescence microscopy as this is a very sensitive method. However, such degradation has never been found.

There is only one exception until now, where the small amounts of fluid which can be obtained from the ectoperitrophic space are sufficient to determine the amount of enzymes: the larva of *Rhynchosciara americana* (Nematocera) which has extremely long and extensive caeca (see Sect. 6.5; Terra et al. 1979; Terra and Ferreira 1981, 1983).

6.4.2 Water Transport Across Peritrophic Membranes

A first attempt to solve this complex problem was made by Zhuzhikov (1964). He investigated mosquito larvae, *Aedes aegypti*. With a tenfold difference of KCl concentration on the two sides of the peritrophic membrane no appreciable membrane potential was created. No selective permeability for simple anions and cations

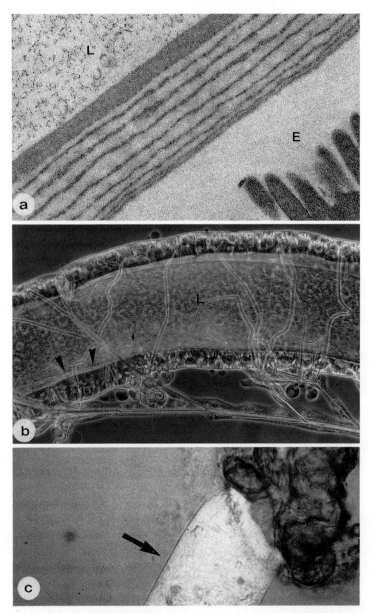

Fig. 71. a Ferritin molecules (364 kDa) do not penetrate the peritrophic membrane of larvae of *Anopheles stephensi* (Diptera, Culicidae) from the lumen side (*L*) to the epithelium side (*E*); 19 200x. *b* FITC-labeled dextran (M$_n$ 2400 daltons) penetrates the single peritrophic membrane (*arrowheads*) of the larva of the blowfly, *Calliphora erythrocephala*. The tube-like peritrophic membrane is crowded on the inside and in the lumen (*L*) with bacteria (see Sect. 6.2). Incident fluorescent light shows the FITC-labeled dextran, whereas phase contrast reveals the outlines of the midgut epithelium and its tracheal supply; 180x. *c* The peritrophic membrane of a *Drosophila melanogaster* larva is filled with ingested fluorescein isothiocyanate-labeled dextran (M$_n$ = 32 000 daltons). The midgut epithelium (*upper right*) has been stripped off to demonstrate that the labeled dextran does not permeate the peritrophic membrane (*arrow*); 300x (After Peters and Wiese 1986)

was observed and the coefficient of permeation of the peritrophic membrane was the same in both directions. The hydraulic conductivity, or as Zhuzhikov called it, the coefficient of filtration, was determined as L_p = "3.36 × 10^{-8} cm^3/sec through an area of 1 cm^2 at a pressure of 1 bar".

This value is contradictory to the value obtained by Zimmermann and Mehlan (1976) for the single peritrophic membrane of the larva and the inner peritrophic membrane of the adult blowfly, *Calliphora erythrocephala*. In spite of the difficulties involved in the determination of the coefficient of filtration and in spite of the dif ferent species used, these authors assumed that Zhuzhikov's value is undoubtedly wrong. They found a quite different order. The L_p of the peritrophic membrane of the blowfly larva was determined to 2.5 × 10^{-2} cm s^{-1} bar^{-1} and that of the adult blowfly 1.4 × 10^{-2} cm s^{-1} bar^{-1} at 20 °C, pressure about 150 mm H_2O, in distilled water as perfusion solution. In Ringer solution the values were higher by about 30%. The L_p increases with increasing pressure and becomes constant above 80 to 120 mm H_2O. From the temperature dependence of L_p the activation energy for the water transport through the membranes of both stages was calculated to be 4.9 and 4.1 kcal/mol, respectively. The half-time of water exchange for peritrophic membranes tied on both ends was calculated from L_p and elastic properties of the membrane to be between 3 and 40 s depending on the applied hydrostatic pressure. The L_p values are of the same order of magnitude as those found for artificial membranes. More experiments and data are needed.

6.4.3. Permeability Properties of Peritrophic Membranes

Investigations on the permeability of peritrophic membranes are restricted mainly to insects. Montalenti (1930, cited in Wigglesworth 1972) investigated termites, *Reticulitermes flavicollis*. Castelnuovo (1934) worked with *Dixippus morosus*, and von Dehn (1933) with honeybees, *Apis mellifica*, and blowflies, *Calliphora erythrocephala*. Only larger colloidal particles such as Prussian blue were retained by the peritrophic membranes, whereas Congo red, Trypan blue, and Aniline blue passed through. It should be considered that only the blowflies form a continuous tube made up of three peritrophic membranes, whereas the other species secrete patches of membranes which altogether form a peritrophic envelope. Of course the continuous membranes of Diptera were investigated preferentially for their permeability properties as all nutritive substances and products of digestion must permeate these membranes in order to reach the midgut epithelium. Schildmacher (1950) used colloidal gold particles and found that particles with a diameter between 2–4 nm passed through the peritrophic membrane of larvae of *Aedes aegypti*. Bowen et al. (1951) found that the single peritrophic membrane of larvae of *Drosophila repleta* is penetrated by ^{140}Ba and ^{140}La.

Wigglesworth reported already in 1929 that the single continuous peritrophic membrane of a tsetse fly, *Glossina* sp., is permeable to hemoglobin (68 kDa). Modespacher et al. (1986) used *G. morsitans*, which already had a fully formed peritrophic membrane (see Sect. 3.1), and fed them ferritin (364/450 kDa) together with human serum. Ferritin remained in the endoperitrophic space and did not penetrate the peritrophic membrane. However, horseradish peroxidase (40 kDa) penetrated the peritrophic membrane. It was transported through intercellular clefts, basal lamina and hemolymph and finally taken up by the fat body. Penetration was observed in the anterior part and to a smaller extent in the middle part of the midgut within 2 h after feeding. An intracellular pathway through the midgut epithelium could not be demonstrated. Nogge and Gianetti (1979) found that a small fraction of albumin (0.035%) taken in by the tsetse fly is transported through peritrophic membrane and midgut epithelium into the hemolymph without previous digestion. In this case FITC-labeled dextran molecules larger than 45 kDa could not pass the peritrophic membrane (Nogge and Peters unpubl. results). Preliminary investigations with fluorescein isothiocyanate-labeled dextrans showed that these dextrans, if fed together with blood, are adsorbed by the blood albumin. Then even the smallest dextran fractions did not permeate the peritrophic membrane. But if instead of blood a glucose solution isosmotic with saline was fed together with ATP as a feeding stimulant, conclusive results could be obtained. In *Glossina palpalis* bovine hemoglobin (68 kDa) did not permeate the peritrophic membrane, whereas myoglobin (16.9/17.5 kDa) could penetrate. Dextrans with an average molecular weight (M_n) of less than 17.5 kDa and an Einstein-Stokes radius (ESR) of 3.22 nm labeled with fluorescein isothiocyanate passed the membrane of *G. palpalis* and *G. morsitans*, whereas dextran with M_n 32 kDa and an ESR of 4.25 nm penetrated only slowly after some time and dextran of 74 kDa did not permeate the membrane. Protein molecules are more densely packed than dextrans. Therefore, albumin with 68 kDa or partially digested albumin might permeate the peritrophic membrane.

Zhuzhikov (1964) investigated the peritrophic membrane system of adult houseflies, *Musca domestica*, consisting of two morphologically different membranes, and of blowflies, *Calliphora erythrocephala*, consisting of three different membranes. Ovalbumin (43.5 kDa), gelatin (37 kDa) and starch did not penetrate these membrane systems in either species. The permeability of starch was tested in both directions. Glucose and maltose permeated readily through the whole system of membranes as well as the single membranes, whereas sucrose was only partially and lactose only weakly permeable. The tested amino acids passed freely through the peritrophic membranes, whereas commercial microbiological peptone solution could not permeate the complete membrane system or the inner membrane which was tested separately. However, the peptone could penetrate the outer membranes weakly. The permeability of amylase was tested using human salivary amylase. For these experiments saliva was transferred to the isolated whole system of peritrophic membranes. The ends of the peritrophic membranes were tied with silk and the sac placed into a microcuvette filled with starch solution at 38 °C. After some time the degree of permeability was determined with the iodine test. Zhuzhikov found that the enzyme cannot permeate from the endoperitrophic space to the exterior through the whole system of membranes, but that it is able to pass readily in the opposite direction, i.e. from outside to inside the membrane system. If the separate layers were tested, these proved to be readily permeable in both directions. Zhuzhikov assumed therefore that the spaces between the membranes are filled with colloidal solutions and that the whole system, including these fluid-filled spaces, behaves differently from its components.

Stable flies, *Stomoxys calcitrans*, were fed on rabbits previously immunized with various fly tissues (Schlein and Lewis 1976). Mortality after 15 days feeding on rabbits immunized with flight muscle proteins was twice that of the control flies fed on an untreated rabbit. Other effects were the paralysis of legs or wings and unequal deposition of the endocuticle. However, it has not been tested whether intact antibodies or only Fc-fragments passed the peritrophic membranes of these flies.

Adult fleshflies, *Sarcophaga falculata*, which have two morphologically different peritrophic membranes, were fed with rabbit immunoglobulins raised against fleshfly tissues like brain, muscle or hemolymph (Schlein et al. 1976). The presence of antibodies in different tissues of the flies could be demonstrated using the indirect fluorescent antibody method. Again, it has not been shown whether intact antibodies or only fragments of antibodies permeated the peritrophic membranes of the flies. Fishman and Zlotkin (1984) investigated the route of transport of horseradish peroxidase (HRP)(M = 40 kDa). HRP readily penetrated the peritrophic membranes and was taken up by the midgut epithelium. Similarly, a polypeptide of the venom of the South African cobra, *Naja mossambica*, which has a low molecular weight of about 7kDa, penetrates easily the peritrophic membranes of these flies (Fishman et al. 1984). This venom polypeptide is paralytic and lethal when taken up orally by the fleshflies. In the figures of both papers only the inner peritrophic membrane is shown, which can be recognized by its electron-dense layer on the lumen side.

The tube-like, continuous peritrophic membranes of Diptera are especially useful to gain insight into the permeability properties of peritrophic membranes. In a comparative investigation Peters and Wiese (1986) used for preliminary experiments several substances of known molecular weight which could be followed by their natural color during their passage through the different parts of the gut. However, increasing dilution made such experiments impossible. Therefore, several other markers like fluorescent dyes, electron-dense markers like hemoglobin and ferritin, and predominantly graded FITC-labeled dextran fractions were used which had been proven already to be useful tracers in studies of glomerular filtration in mammals (Renkin and Gilmore 1973; Caulfield and Farquhar 1974).

The results of these investigations are compiled in Table 15. It should be considered that charged protein molecules are more compact and smaller than dextran molecules which occur in solution as random coils. Therefore, the effective radii and not the molecular weight have to be compared in connection with permeability investigations using both types of substances. The fluorescent dyes FITC (M = 389.4 daltons) and Evans blue (M = 980.8 daltons) were able to penetrate the peritrophic membrane of all species tested. Ferritin was fed to smaller species such as the larvae of *Anopheles stephensi*, *Anisopus cinctus* and *Chironomus thummi*. But this large molecule did not penetrate the peritrophic membrane. Horseradish peroxidase (40 kDa) could be demonstrated only on the lumen side of the inner peritrophic membrane of adult blowflies, *Calliphora erythrocephala* where it accumulated. This proved to be due to the presence of mannose-specific lectins (see Sects. 5.3 and 6.2).

These comparative investigations give some hints concerning the properties of peritrophic membranes which may be responsible for the differences in permeability. First, the texture of microfibrils should be discussed. The membranes of the different species show either a random or an orthogonal texture of microfibrils. However, if an orthogonal texture is present, the spacings cannot be considered as a limiting

Table 15. Permeability of peritrophic membranes of Diptera to Evans blue (first column) and dextrans labeled with fluorescein isothiocyanate (FITC) following columns (After Peters and Wiese 1986)[a]

$M_n^b \times 10^{-3}$		2.3	4–6	6.2	17.2	32	37.2	74
ESR[c]		1.3	1.7	2	3.2	4.3	4.5	4.2
Larvae								
Aedes aegypti	+/–	–	–	–	–	–	–	–
Anopheles stephensi	+	+	+	+	+	–	–	–
Culex pipiens	+	+	+	+	+	+	–	–
Odagmia ornata	+	+	+	+	+	+	–	–
Anisopus cinctus	+	+	+	+	+	+/–	–	–
Sarcophaga barbata	+	+	+	+	+	+/–	–	–
Calliphora erythrocephala	+	+	+	+	+	+/–	–	–
Adults								
Sarcophaga barbata	+	+	+/–	–	–	–	–	–
Calliphora erythrocephala	+	+	+	+/–	–	–	–	–

[a] +, Readily permeable; +/–, slowly penetrating; –, impermeable.
[b] M_n, Number average molecular weight.
[c] ESR, Einstein-Stokes radius.

factor because the spacings amount to about 100–200 nm in larval peritrophic membranes of *Odagmia ornata* (Simuliidae) (Fig. 46) which means that even the largest molecules, e.g. Blue dextran, should be able to pass freely through these highly ordered grids. The network of microfibrils seems to be of importance with respect to the considerable tensile strength of these membranes.

Proteoglycans (acid mucopolysaccharides), which have been demonstrated in different amounts and patterns, might be responsible for the different permeability properties of these membranes. Especially in the peritrophic membranes of larvae of Nematocera these substances appear in an unexpected variety of fine structural patterns (Figs. 39–46). These patterns may differ between families and even genera of a given family (Peters 1976, 1979). They have been confused with the chitin moiety by Richards and Richards (1971, 1977). A thick layer of electron-dense proteoglycan-containing material is present in similar thickness on the lumen side of the peritrophic membrane of the larvae of *Aedes aegypti* (Fig. 44) and *Anopheles stephensi* (Fig. 71; Culicidae). Moreover, a different number of additional layers of similar appearance but with irregular holes appear inside the membrane. However, in sharp contrast to these morphological similarities, the permeability properties of these membranes are quite different. The peritrophic membrane of *Aedes aegypti* larvae is impermeable to the smallest fraction of FITC-labeled dextran with a molecular weight (M_n) of 2.4 kDa and an Einstein-Stokes radius of 1.3 nm. The peritrophic membrane of *Anopheles stephensi* is still permeable to labeled dextrans with an M_n of 17.5 kDa and an Einstein-Stokes radius of 3.2 nm.

These results are contradictory to those of Zhuzhikov (1970, 1971) who claimed that colloidal gold particles up to a diameter of 9 nm were able to penetrate the peritrophic membrane of *Aedes aegypti* larvae, which means a tenfold greater size. He obtained his colloidal gold solution by the reduction of "hydrochloroauric acid" with white phosphorus and added powdered "Mipor" as an inert solid substance together with the mosquito larvae. It is still unclear how the unstable colloid could withstand these conditions without aggregation and precipitation. Furthermore, Zhuzhikov (1971) calculated an effective radius from the reflexive coefficient and found that it varies for smaller and larger molecules: 4000 nm for sucrose, 300 nm for raffinose and only 25 nm for inulin. Therefore, he suggested that the permeability of nonionic solutions not only depends on the pore size but might be regulated by an interaction between the matrix of the peritrophic membrane and the permeating molecules, the interaction being stronger with increasing size of the molecules.

High molecular aggregates and colloidal food substances cannot pass the peritrophic membrane. Peters and Wiese (1986) found that ferritin did not penetrate the peritrophic membrane of larvae of *Anopheles stephensi* (Fig. 71a), *Chironomus thummi* and *Anisopus cinctus*.

The peritrophic membrane of the larvae of another species of the Culicidae, *Culex pipiens fatigans* (Fig. 44), has an electron-dense layer on the lumen side which is characterized by 15-nm pores in high density. In this case even dextrans with M_n of 32 kDa and an Einstein-Stokes radius of 4.3 nm are able to permeate the membrane. These pores are much larger than the permeating molecules.

On the lumen side of the peritrophic membrane of the blackfly larva, *Odagmia ornata*, two layers of electron-dense rods are arranged in orthogonal patterns (Fig. 46). The spacings between the compact rods of the second layer are about 20 nm. The labeled dextran with an M_n of 32 kDa readily permeates this membrane. However, the next larger fraction with an M_n of 37.2 kDa and an only slightly larger Einstein-Stokes radius of 4.5 nm did not penetrate it.

Obviously, the assumption that the holes filled with electron-lucent material might be the pathway for permeating molecules appears to be wrong. The peritrohic membranes of other species which have a continuous, electron-dense and proteoglycan-containing layer on the lumen side have similar or quite different permeability properties despite their similar, fine structural appearance (see Table 15 and for comparison the structural details in Figs. 39–46, 50, 59–63).

It should be emphasized that the permeability of larval and adult membranes may differ. Nothing can be said so far about the importance of two or even three morphologically as well as biochemically different peritrophic membranes in adult flies, e.g. *Sarcophaga barbata* or *Calliphora erythrocephala* (Figs. 50, 61, 62).

An interesting problem is the permeability of the peritrophic membrane to proteolytic enzymes which are secreted by the midgut epithelium. Detra and Romoser (1979) investigated this problem with a simple method in vitro. The peritrophic membrane was dissected, ligated on both ends and covered with molten agar at approximately 43 °C. Azure-blue hide powder (Calbiochem) was used as an indicator of proteolytic activity. Capillary tubes with buffer and crystals of azure-blue hide powder were positioned at defined distances from the peritrophic sac. Azure-blue hide powder is a dye-protein complex which is insoluble in water. As soon as the protein component is digested by proteases, the water-soluble dye is released into the buffer solution (Rinderknecht et al. 1968). It appears that the peritrophic membrane is permeable to the proteolytic enzyme of the *Aedes aegypti* larva.

A more sophisticated method was used by Terra and Ferreira (1983). These authors and their coworkers determined the distribution and size of digestive enzymes in the midgut epithelium as well as the ecto- and endoperitrophic space with biochemical methods in three different species. Thus, the effective pore size of the peritrophic membrane of the larvae of two species of Diptera of the family Sciaridae, *Rhynchosciara americana* and *Trichosia pubescens*, was determined to be 7.5–8 nm. In this case the peritrophic membrane is secreted by the cardia as a single, continuous tube. The effective pore size of the peritrophic envelope of the larva of the cassava hornworm, *Erinnyis ello*, belonging to the Lepidoptera Sphingidae, was determined as 7–7.5 nm (Santos and Terra 1986). In the latter species the peritrophic envelope consists of numerous peritrophic membranes which are secreted by the whole midgut epithelium.

For the first time small Crustacea, *Daphnia pulex*, were investigated by Hansen and Peters (unpubl. results). *Daphnia* was fed FITC-labeled dextrans of different sizes, Dextran blue with M 2000 kDa and latex particles of different sizes. All dextran fractions tested were able to penetrate the peritrophic envelope consisting of several peritrophic membranes. Latex beads with a diameter of 130 nm permeated also, but latex particles with a diameter of 327 nm could not permeate (Fig. 70). This shows that even a loosely packed group of peritrophic membranes does not permit the penetration of all particles, but that it has a certain threshold. The effective pore size could be in the range of 300 nm.

In graminivorous locusts tannic acid is able to penetrate the peritrophic envelope, whereas in polyphagous species tannin does not permeate the peritrophic membranes but is adsorbed to them and voided with the fecal pellets (Bernays and Chamberlain 1980; Bernays et al. 1980; see Sect. 3.3).

The permeability of the peritrophic envelope of fifth instar larvae of the tobacco hornworm, *Manduca sexta*, which is composed of many peritrophic membranes, was investigated by Wolfersberger et al. (1986). The peritrophic envelope was isolated, rinsed carefully, filled with a polypeptide solution and ligated to form a closed sac. It was rinsed and placed in an isotonic salt solution. Aliquots of this solution were assayed at regular intervals for protein. The half-time for diffusion increased linearly with the log of molecular weight up to 66 kDa. Larger proteins needed more time for diffusion. Proteins up to a molecular weight of 100 kDa were able to permeate through the peritrophic envelope of this species. Blue Dextran, with a molecular weight of about 2000 kDa, did not penetrate the membranes.

Adang and Spence (1983) investigated in vitro the permeability of the peritrophic envelope of late instar larvae of the Douglas fir tussock moth, *Orgyia pseudotsugata*. Pieces of the isolated peritrophic envelope, consisting of many peritrophic membranes, were mounted in a small chamber and tested with ^{14}C-glucose, polyethylene glycol 6000, 300-nm latex beads and 1.1-µm polystyrene latex beads, bacteriophages ϕX-174 from *Escherichia coli*, *Bacillus thuringiensis* strain HD-1 and its spores. Extrapolation shows that the peritrophic envelope has a maximum effective pore size of about 650 nm. Treatment of the membranes with ficin or the dissociated entomocidal parasporal crystal of *B. thuringiensis* had no effect on their permeability properties.

Recent investigations on the effects of the δ-endotoxin of *Bacillus thuringiensis* var. *entomocidus* on the larval midgut of *Spodoptera littoralis* (Lepidoptera, Noctuidae) proved that peritrophic membranes play an unexpected part in its mode of action (Yunovitz et al. 1986). *Bacillus thuringiensis* is used as an insecticide as it produces a parasporal crystal containing an endotoxin. The midgut epithelium is the first interaction site for the endotoxin. An isolated fraction with M_r 64 kDa was toxic to isolated midgut cells and induced cell lysis. However, if detergents were removed by dialysis, the toxic protein fraction formed aggregates of high molecular weight which were unable to penetrate the peritrophic membranes and therefore did not reach the sensitive midgut cells. Cell lysis was not due to the detergents as could be shown in experiments using detergents in the absence of toxin. In vivo proteases seem to degrade the high molecular form of the endotoxin into smaller molecules which can penetrate the peritrophic envelope and destroy the midgut cells.

Recently, data have been obtained on the permeability of the peritrophic envelope of a mosquito, *Culex nigripalpus* (Borovsky 1986). After a female mosquito has taken a blood meal several important biological processes are induced consecutively: the synthesis and secretion of a peritrophic envelope consisting of many peritrophic membranes, the synthesis and secretion of proteolytic enzymes, vitellogenesis and the formation of eggs. A different time is required for the secretion of the peritrophic envelope (see Sect. 3.3.2). In *C. nigripalpus* 24 h are needed. Subsequently, protease activity increases and reaches a peak 35 h after the uptake of blood. During proteolysis of the food bolus proteins, degradation products of different size could be found. Again, the peritrophic envelope acts as an ultrafilter. A larger protein, similar to midgut protease, migrated through the peritrophic membranes into the blood bolus; however, the quantity of these proteins was at least tenfold higher than trypsin. Amino acids, peptides and polypeptides penetrated the peritrophic membranes and reached the ectoperitrophic space. The largest polypeptides in this space had an average molecular weight of 23 ± 2 kDa. They were further digested and adsorbed by the midgut epithelium.

For comparison of these permeability data, only results obtained with glomerular or podocyte filtration can be used. It has been shown by physiological experiments that in man, dog, cat, rabbit and rat proteins, which are more compact than dextrans, do not penetrate the basal lamina of the podocytes of the glomerulus in the kidney if their molecular weight exceeds 68 kDa (Renkin and Gilmore 1973). Caulfield and Farquhar (1974) investigated electron microscopically the permeability of the glomerular basal lamina of the rat for dextran fractions. These authors could show that dextran fractions with molecular weights greater than 62 kDa and up to 125 kDa were retained more and more. Landis and Pappenheimer (1963) found that in muscle capillaries of rats two types of small pores seem to exist, round pores with a diameter of about 5 nm and slits of about 5 nm in width.

Investigations in invertebrates are rather scanty. In capillaries of an unidentified *Octopus* species, Browning (1979) observed that the large ferritin molecule was retarded but could still permeate. The basal lamina of the podocytes of the antennal gland of the crayfish retained proteins between the size of albumin (69 kDa) and mammalian serum globulin (150 kDa) (Kirschner and Wagner 1965).

6.4.4 Use of Peritrophic Membranes for Cellular Sieving

Human blood platelets are highly sensitive to the forces which occur during separation from plasma. Centrifugation and other separation methods tend to induce platelet aggregation or cause damage that leads to activation. Man-made filters are not optimal as their relatively low porosity means that substantial pressure has to be exerted for sieving platelets from plasma. Therefore, Skaer (1981) suggested the use of peritrophic membranes from cockroaches (*Periplaneta americana*) or locusts (*Locusta migratoria*). Their pore size is considerably smaller than that of platelets and their high porosity allows the use of high fluxes and therefore a rapid and effective separation of platelets from serum in quantities which are adequate for many experimental purposes. The effective pore size of the peritrophic membranes of *P. americana* is in the range 125 and 150 nm. The peritrophic envelope has to be isolated, cleaned and sandwiched between two Teflon washers in a modified, small, stainless-steel Millipore filter holder. Platelet-rich plasma is sieved under the influence of gravity, without using pressure, until the membrane becomes clogged with platelets and white cells. Then the direction of the flow is reversed momentarily in order to declog the membranes. The operation should be repeated several times. Thus, 1 ml of plasma can be sieved within 5 min.

6.5 Compartmentalization of the Midgut by Peritrophic Membranes

Obviously, only a few authors regarded the compartmentalization of the midgut lumen by peritrophic membranes. The endoperitrophic space contains food and food residues which are degraded by different enzymes, whereas the ectoperitrophic space is filled with other enzymes and degraded nutrients which permeated through the peritrophic envelope. The latter are degraded further before they are absorbed by the midgut epithelium. Due to the fact that the ectoperitrophic space is normally extremely small and inaccessible to physiological measurements, most authors, especially physiologists, disregarded these compartments, e.g. Treherne (1967), Berridge (1970), and Dow (1981) who studied water and ion movements as well as nutrient absorption in the locust, *Schistocerca gregaria*, and Reynolds et al. (1985), who investigated the food and water economy of larvae of the tobacco hornworm, *Manduca sexta*. In the preceding section is has been shown that, if these membranes form continuous tubes surrounding food and food residues, all nutrients must permeate a peritrophic envelope which may have limited and perhaps even selective permeability before they can reach the midgut epithelium.

6.5.1 Countercurrent Flow in the Midgut and the Role of the Midgut Caeca in Locusts

Dow (1981) administered the dye amaranth as a marker either together with food or injected into the hemolymph of locusts, *Schistocerca gregaria*. If the locusts were fed ad libitum, the dye moved gradually posteriorly together with the food. However, in animals deprived of food for 2–4 h prior to an experiment, the dye accumulated in the gastric caeca. Although Dow did not recognize the presence of a peritrophic envelope made up of many peritrophic membranes surrounding the food, he assumed that fluid moves anteriorly between food and midgut epithelium as the dye accumulates in the caeca. Dow supposed that under these circumstances a water flow occurs from the lumen of the gastric caeca into the hemolymph which should be driven by an ion transport coupled with it. In the posterior part of the midgut, fluid might be replaced partially by the Malpighian tubules. Thus, a countercurrent flow could be possible in the midgut which might allow a better use of nutrients and enzymes during intervals of hunger. In locusts a special problem is the elimination of ingested tannin and plant toxins. Experimental evidence based on the absorption of dyes suggested that the presence of the pockets in S. gregaria (see Sect. 3.3.2) is related to rapid water absorption and perhaps to the sequestration of plant toxins ingested in the very polyphagous diet of this species Bernays (1981). The effects of tannin on the growth and development of 15 species of Acridoidea were investigated by Bernays et al. (1980). Polyphagous species showed no deleterious effects. Tannin proved to be unable to penetrate the peritrophic membranes. It was hydrolyzed at least partly to tannic acid and glucose (Bernays and Chamberlain 1980) or remained adsorbed onto the peritrophic membranes during their movement through the gut and was eventually discharged with the fecal pellets. More than 30% of the fecal tannin was associated with the peritrophic membranes investing the remains of food. In graminivorous species reduced growth rates and poor survival were observed. In these species tannin could penetrate the peritrophic membranes, damaged the midgut epithelium and reached the hemocoel.

6.5.2 Countercurrent Flow in the Midgut and the Role of the Midgut Caeca of Mosquito Larvae

Wigglesworth (1933b) and Ramsay (1950) fed light microscopically detectable dyes to mosquito larvae. They observed after some time a concentration of dye in the caeca and suggested therefore an anteriorly directed movement of fluid in the midgut. This accumulation of dyes in the caeca led to the assumption of a coupled ion and water flux in the epithelium of the caeca which could cause an anteriorly directed movement of fluid within the ectoperitrophic space. Larvae which had been adapted to salt water were able to absorb components of the midgut fluid with their caeca, whereas larvae which had been reared in freshwater only, and then transferred to seawater, were not (Wigglesworth 1933b). Asakura (1982) could demonstrate by light microscopic techniques that in larvae of *Aedes togoi*, living in seawa ter, some cells of the caeca contain considerable amounts of chloride ions. The accumulation of chloride ions increased in larvae which were reared in double concentrated seawater. However, if larvae of the same species were adapted to freshwater, only small concentrations of chloride were found. These results support the suggestion that the caeca of mosquito larvae are involved in osmoregulation.

Volkmann and Peters (1989b) reinvestigated the functions of the caeca of mosquito larvae using fluorescing dyes, fluorescein isothiocyanate (FITC, M = 389.4) and Evans blue (M = 960.8), which are detectable even in extremely low concentrations, and which have apparently no toxic effects. The following results were obtained with fourth instar larvae of *Aedes aegypti* and *Culex pipiens*. Due to an irregular uptake of food, the feeding time had to be varied. Some larvae had already taken up enough dye together with the food after 2 min of feeding, whereas others needed more than 30 min. Both dyes were able to penetrate the peritrophic membrane. Soon after feeding they appeared in the ectoperitrophic space and in the caeca. However, usually the caecal lumina were filled only partially. In larvae of *Aedes aegypti* the dyes appeared only in the anterior part. This part of the caeca is separated from the remaining part by the caecal membrane (see Sect. 3.3.2). In *Culex pipiens* approximately two-thirds of the caeca was filled with accumulating dye and only the hindpart remained clear which is separated by the caecum membrane. Therefore, it can be concluded that the dyes were unable to penetrate the caecal membrane.

Movement of fluid in the ectoperitrophic space could be observed only if FITC was present in the culture medium in very low concentrations of < 0.1 mM. The fluorescent dye accumulated in the ectoperitrophic space in the region of the third to fifth abdominal segment. The fluid was pumped into the anterior part of the midgut and into the caeca by antiperistaltic movements of the midgut until no fluorescence was observable in the original position, the ectoperitrophic space. Then the eight caeca contracted synchronously and pressed the fluorescing fluid back into the midgut. These movements were repeated in irregular intervals. After some hours the fluorescence diminished in the ectoperitrophic space but was still detectable in the caeca. However, after 2 days only small amounts of FITC could be observed in the caeca. Comparable results were obtained with larvae which had been starved for 9 days, as well as those which were reared in distilled water or in 10 or 20% artificial seawater.

Thus, it could be shown that antiperistaltic movements of the midgut are mainly responsible for the countercurrent flow in the ectoperitrophic space, and not only the drag induced by the assumed, combined ion and water flux through the epithelium of the caeca into the hemolymph. This can be observed in a transparent mosquito larva, but not in a locust.

The caecal membrane forms a permeability barrier (Figs. 32–34). Its morphological and permeability properties differ from those of the peritrophic membrane. Fluorescent dyes such as FITC and Evans blue penetrate the peritrophic membrane (Peters and Wiese 1986; see Sect. 6.4) but not the intact caecal membrane. Therefore, these dyes accumulate in front of the caecal membrane, whereas cellular debris accumulates behind and on the newly formed caecal membrane. These investigations suggest the following functions of the caeca of mosquito larvae. Antiperistaltic movements of the midgut cause a counterflow which provides the caeca with midgut fluid. Water, nutrients and ions are resorbed and transferred to the hemolymph by specialized cells, the ion-transporting cells. According to the salinity of the breeding medium, the ion-transporting cells alter their physiological state which can be detected by ultrastructural changes. Substances which cannot be resorbed accumulate in the lumina of the caeca. They are retained by the caecal membrane in the anterior part of the caeca.

In order to obtain information on the role of caecal cells in osmoregulation, larvae were reared in water of different salinities: in distilled water, freshwater, or artificial seawater diluted to 10 or 20% salt. No structural changes in the caeca could be observed light microscopically. However, in larvae reared in distilled water electron microscopy revealed that in resorbing/secreting cells the basal labyrinth was enlarged considerably. The channels and the intercellular spaces were widened. In the apical cytoplasm the number of mitochondria increased as compared with those of larvae reared in freshwater. The number and distribution of other cellular components remained unchanged. In ion-transporting cells the basal labyrinth did not change remarkably, but the number of mitochondria entering the microvilli increased considerably (Fig. 72). In *C. pipiens* all microvilli contained at least one mitochondrium, but often two or even more (Fig. 72). In larvae reared in 10% seawater the basal labyrith of the resorbing/secreting cells was developed in *Ae. aegypti* and *An. stephensi* as in larvae reared in freshwater, but in *C. pipiens* it was more elaborate without widened intercellular spaces. The number of mitochondria and the length of microvilli appeared to be slightly reduced. The number of mitochondria in the microvilli and the length and diameter of microvilli of ion-transporting cells were reduced as compared with larvae grown in freshwater. However, microvilli lacking a mitochondrion still contained portasomes, smooth endoplasmic reticulum, free ribosomes and microfilaments. Portasomes were defined by Harvey (1980) as small globular particles which have a diameter of 14 nm. They are located on the cytoplasmic side of the microvillar membrane of ion-transporting cells. They are supposed to be K^+-ATPases. Larvae grown in 20% seawater showed in both cell types a considerable reduction in the size and the length and diameter of their microvilli, the mitochondria seemed to be dislocated from the apical into the basal region, and sometimes the channels of the basal labyrinth reached the apical part of the cells, but the intercellular spaces were not enlarged. The size of the ion-transporting cells of the three species was distinctly reduced just as the length of their microvilli. Mitochondria could be found in the microvilli only occasionally in *Ae. aegypti* and *An. stephensi*, but not at all in *C. pipiens* (Fig. 73). The microvilli still contained, however, portasomes, microfilaments, free ribosomes and smooth endoplasmic reticulum. Light and electron microscopic observations using Intense-silver enhancer (Janssen, Belgium) or Komnick's silver lactate method (1969; see Geyer 1973) revealed that all caecal cells are able to accumulate chloride ions, but to different degrees. Considerable amounts of chloride ions were restricted to the ion-transporting cells. A strong reaction could be observed in the brush border of these cells, whereas the cytoplasm and the basal labyrinth contained less chloride ions. After rearing the larvae in water of different salinities, it was revealed that the number of cells accumulating chloride ions changed with the respective external ion concentration. The number of "chloride cells" increased in larvae reared in 10 or 20% seawater, whereas in larvae adapted to freshwater or distilled water, these cells were less numerous. In the caeca of larvae reared in water containing low concentrations of salt, mostly the ion-transporting cells reacted

171

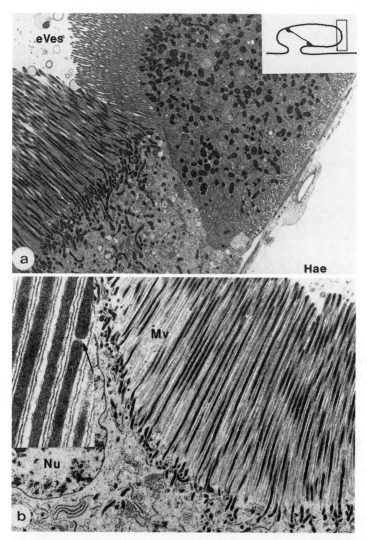

Fig. 72. Culex pipiens. a An ion-transporting cell with mitochondria in its microvilli and a resorbing/secreting cell with electron-dense plasma and an elaborate basal labyrinth. *HAE* hemolymph; *eVES* extracellular vesicles; 2900x. *b* An ion-transporting cell of a larva which had been reared in distilled water; 5100x. *Inset:* Sometimes two mitochondria (*arrow*) occur in a single microvillus; 26 000x (After Volkmann and Peters 1989b)

positively to the chloride ion assay, whereas the number of resorbing/secreting cells accumulating chloride ions increased with external salinity. If the results of electron microscopic and physiological investigations on osmoregulation in insects (Bradley 1987) are applied to the similar results observed in the cells of the caeca of mosquito larvae, it can be suggested that the ion flux is directed from the lumen of the caeca to the hemolymph. Mosquito larvae which live in freshwater must be able to

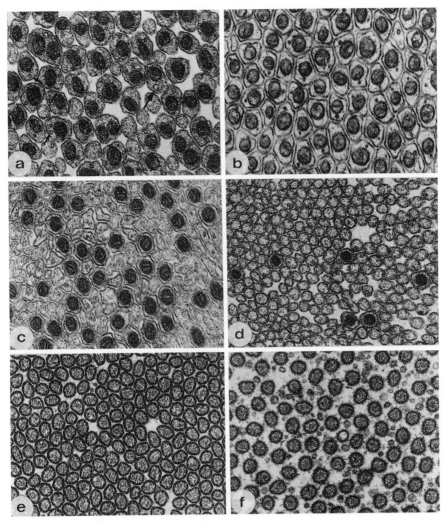

Fig. 73a-f. Cross-sections through the microvilli of ion-transporting cells of larvae which were reared under different osmotic conditions (after Volkmann and Peters 1989b). *a* In *Culex pipiens* larvae grown in freshwater the microvilli contain 1–2 mitochondria, smooth endoplasmic reticulum (*small arrows*), ribosomes and microfilaments; 23 200x. *b* in *Culex pipiens* larvae reared in distilled water nearly all microvilli contain mitochondria; 15 900x. *c* Only a few microvilli of ion-transporting cells of *Aedes aegypti* larvae grown in 10% seawater contain mitochondria; 15 900x. *d* In larvae of *Aedes aegypti* grown in 20% seawater the number of mitochondria in the microvilli is obviously reduced; 18 000x. *e* In *Culex pipiens* larvae reared in 20% seawater no mitochondria appear in the microvilli of ion-transporting cells; 35 000x. *f* In *Culex pipiens* larvae grown in 20% seawater the microvilli of the resorbing/secreting cell look different from those of an ion-transporting cells as compared with *e*; 28 100x

minimize the uptake of water and the leakage of ions into the medium. This is achieved by reducing the drinking of water, by the production of diluted urine in order to get rid of water (Ramsay 1950), by the reabsorption of ions in the hindgut (Ramsay 1950), by the absorption of ions against a concentration gradient by the anal papillae (Koch 1938; Stobbart 1971), and by the recycling of ions which are transported through the caecal epithelium into the hemolymph of mosquito larvae (Volkmann and Peters 1989b).

6.5.3 Compartmentalization of the Midgut and Diversity of Enzymes

Larvae of *Rhynchosciara americana* have extremely long caeca which allow the investigation of their digestive enzyme content (Ferreira and Terra 1982). Amino- and carboxypeptidases could be localized in the plasma membrane and cytoplasm of the caecal cells. Therefore, these authors assumed that in this species a considerable part of digestive processes takes place at the surface of the caecal epithelium. The results presented by Volkmann and Peters (1989b) suggest that in mosquito larvae similar processes take place.

Terra et al. (1979) assumed that soluble proteases are lost quickly with the feces, whereas enzymes in the ectoperitrophic space, between peritrophic membrane and midgut epithelium, are conserved and used for the second step in the breakdown of nutrients. However, this can only be achieved with continuous tube-like peritrophic membranes as in Diptera. Terra et al. (1979) and Ferreira and Terra (1980, 1982) investigated the distribution of enzymes in the ecto- and endoperitrotphic space as well as the midgut cells of the larva of *Rhynchosciara americana* (Diptera, Sciaridae). Normally, it is almost impossible to obtain enough fluid for analyses from the ectoperitrophic space. But this species proved to be particularly suitable because it has a single, tube-like, continuous peritrophic membrane and large midgut caeca which allow one to determine the enzyme content of the ectoperitrophic space. Amylase, cellulase and proteinases were present in midgut cells as well as in the ecto- and endoperitrophic space. Disaccharidases (except trehalase), carboxipeptidases, dipeptidases, carboxylesterases and phosphatases could be demonstrated only in the midgut cells. The degradation of food obviously occurs stepwise in these different compartments: in the endoperitrophic space, inside the peritrophic membrane, the breakdown of larger food molecules takes place. The next step, the hydrolysis of these products into oligomers and dimers, occurs in the ectoperitrophic space. Finally, these are degraded in the midgut cells. The authors assume that two different reaction spaces are formed by the presence of a peritrophic membrane. This results in the conservation of enzymes in the ectoperitrophic space, whereas those in the endoperitrophic space are lost quickly with the feces. Terra and Ferreira (1981) assumed a mechanism by which enzymes are recovered from the undigested food before it is excreted. Trypsin, amylase and cellulase are able to penetrate the peritrophic membrane of *Rhynchosciara americana* in order to change from the endo- to the ectoperitrophic space. This seems to occur perhaps in the posterior midgut where the polymeric food molecules are hydrolyzed to sufficiently smaller molecules which might be able to penetrate the peritrophic membrane together with or independently of the enzymes. The authors suggested that a water flux from the hemolymph into the posterior midgut and from there into the anterior midgut and the caeca occurs which then goes back into the hemolymph. This water flux could displace the enzymes from the ectoperitrophic space of the posterior hindgut into the caeca (Ferreira et al. 1981).

The results of the investigations on the distribution and characterization of the digestive enzymes of the larvae of the cassava hornworm, *Erinnyis ello* (Lepidoptera), obtained by Santos and Terra (1986) are simplified and compiled in Table 16.

β-N-acetylglucosaminidase and carboxypeptidase A are probably secreted into the ectoperitrophic space, but they are prevented from passing through the peritrophic envelope. However, β-N-acetylglucosaminidase also occurs in major amounts as a soluble enzyme in midgut cells. α-Galactosidase, β-glucosidase and trehalase are not secreted into the ectoperitrophic space but remain membrane-bound in the midgut cells.

Amylase and trypsin are the smallest enzymes of *Erinnyis ello* larvae. Therefore, they are able to penetrate the peritrophic envelope and to reach the endoperitrophic space. Santos et al. (1986) proposed the following hypothesis: these enzymes are at first integral proteins of the membrane of small vesicles which occur in the midgut cells. Then the enzymes are processed and become soluble inside these vesicles. Part of the enzymes remains membrane-bound, whereas the main part is set free into the midgut lumen. Due to their small size, they are able to pass through the peritrophic envelope and therefore occur in the endoperitrophic space.

Terra et al. (1985) proposed that holometabolous ancestors secreted oligomer and dimer enzymes into the midgut. These small molecules with a diameter below 7.5 nm were able to pass through peritrophic membranes. Therefore, a circulation of enzymes from the ecto- to the endoperitrophic space was possible, which could be driven by secretion of fluid in the posterior region of the midgut and absorption of fluid in the caeca situated at the entrance of the midgut. According to this hypothesis a change in size of the hydrolases occurred during evolution of panorpoid

Table 16. Occurrence of enzymes in the ecto- and endoperitrophic space, their relative molecular weight (M_r) and the diameter of the hydrated enzymes (nm) (After Santos and Terra 1986)

Enzyme	Ecto- Endo-peritrophic space (mU/mg protein)		M_r[a] Centrifugation	Electrophoresis	Diameter[b] (nm)
β-N-acetylglucosaminidase	2.8	0.41	134	149	≈ 8.6
α-Galactosidase	0.45	0.13	81	39	7.3/5.7
Trehalase	16.1	7.1	106	100	≈ 7.9
Amylase (soluble)	970	1400	56	40	≈ 6.5
Trypsin (soluble)	10	10	61	48	≈ 6.5
Carboxypeptidase A	50	17	90	17	7.3/2.3
Aminopeptidase			122		

[a]M_r values were calculated from ultracentrifugation and electrophoretic data.
[b]Diameters of the hydrated enzymes were interpolated in a plot of log (M_r) against Stokes radius for 11 proteins (for details, see Terra and Ferreira 1983 and Santos and Terra 1986).

ancestors which resulted in the formation of larger oligomeric enzymes, with a diameter of more than 7.5 nm, and consisting of several identical or heterogeneous subunits. These were unable to penetrate the peritrophic membranes and therefore were restricted to the ectoperitrophic space or remained membrane-bound in the midgut cells or as a component of the surface coat of these cells. In *Erinnyis ello* larvae, the α-galactosidase proved to be a dimer composed of two identical subunits with M_r 39 kDa, and the carboxypeptidase A is a hexamer composed of three different subunits with M_r 17 kDa.

Peritrophic membranes might play an efficient but as yet unidentified role in digestion which requires further investigation.

6.6 Peritrophic Membranes as a Barrier to Parasites

Parasites which develop and multiply in the midgut epithelium or in the hemolymph must penetrate the peritrophic membranes. This seems to be easy in the case of peritrophic envelopes consisting of loosely packed peritrophic membranes. It is also possible during ecdysis and in the case of blood-sucking mosquitoes a few hours after a blood meal when the secretion of material for the peritrophic membranes has already started but has not yet resulted in the completion of these membranes. The continuous tube-like peritrophic membranes of the Dipteran type were believed to be impermeable to parasites. It was assumed that parasites had either to circumvent the membrane in the hindgut or to use holes in these membranes, regardless of the origin of such holes. However, such mechanisms do not seem to exist. Some pathogenic microorganisms seem to be able to penetrate these membranes and larger ones were observed even to rupture peritrophic membranes. It should be noted that even partial protection from parasites would help to reduce the danger of a massive infection which could hamper or even kill the insect vector, for instance by the destruction of its flight muscles. In the following, examples from different groups of parasites will be described. They show that our knowledge is rather scanty. Discussions of this problem can be found in Orihel (1975), Maier (1976) and Bignell (1981).

The coordination of the time frame of formation of peritrophic membranes and penetration of parasites is of great importance. Those parasites which have no development in the midgut, e.g. microfilariae, try to leave the midgut as soon as they arrive in this region of the gut. However, parasites like Haemosporidia, which have to develop through gamogony, must reach the stage of a mature ookinete with a complete pellicle as soon as possible, in fact in about 24–48 h, to then penetrate the peritrophic envelope and the midgut epithelium. The pellicle is needed to withstand at least for a short time the action of proteases. The Piroplasmida, probably the sister group of the Haemosporidia (Mehlhorn et al. 1980), need much more time to complete their gamogony than the Haemosporidia, about 10–14 days. It is possible therefore that they use ticks instead of mosquitoes as vectors.

As a result of probably long-time selection, the pathogenic parasites are able to penetrate the peritrophic membranes. However, no one has regarded related species which are either poor vectors or are unable to transmit such parasites. For instance, it is still unknown why tsetse flies or blood-sucking bugs do not transmit micro-filariae (Lavoipierre 1958) or Piroplasmida. It is remarkable that in some vectors, e.g. lice and adult fleas, peritrophic membranes are lacking.

6.6.1. Viruses

A nuclear polyhedrosis virus is able to penetrate the loosely packed peritrophic membranes in the midgut of the armyworm, *Pseudaletia unipuncta* (Tanada et al. 1975). They reach the midgut epithelium about 4 h after infection. The membranes of the polyhedra are digested and their virus bundles released. Nucleocapsids as well as virus bundles attach to the microvilli of the midgut epithelium and subsequently penetrate the plasma membrane. This could be enhanced by an enzyme from a granulosis virus.

The larvae of two other species of Lepidoptera, *Trichoplusia ni* and *Spodoptera frugiperda* could be infected with granulosis and nuclear polyhedrosis viruses because these viruses readily penetrated the peritrophic envelope, consisting of patch-like peritrophic membranes, and could be observed in the ectoperitrophic space (Stoltz, Summers and Kawanishi, unpubl. data).

The polyhedral body of the nuclear polyhedrosis virus of the Douglas fir tussock moth, *Orgyia pseudotsugata*, which has an approximate diameter of 2 μm, is unable to penetrate the peritrophic envelope of the larvae of this species. However, after its dissolution the 125 × 60 nm measuring bundles of nucleocapsids of this virus could pass through it (Adang and Spence 1983).

However, the single, continuous, tube-like peritrophic membrane of mosquito larvae seems to be an effective barrier to the passage of viruses. This could be shown by the investigations of Stoltz and Summers (1971) on the pathway of infection of mosquito iridescent virus. First- and second-instar *Aedes taeniorhynchus* larvae could ingest large amounts of this virus without causing a high rate of infection. Most, if not all, ingested virus particles proved to be degraded shortly after entering the midgut. Moreover, granulosis virus capsules from *Trichoplusia ni*, nuclear polyhydrosis virus from *Spodoptera frugiperda*, and *Tipula* iridescent virus from *Tipula paludosa* were also unable to penetrate the peritrophic membrane (Fig. 74). Therefore, infection of larvae of mosquitoes and other Diptera seems to be more likely through cuticle and hemocoele than across the peritrophic membrane and midgut epithelium.

In adult mosquitoes the secretion of peritrophic membranes is induced by the ingestion of a blood meal. Therefore, viruses have a chance to invade the midgut epithelium during the first 4 h after feeding (Houk et al. 1979). After that time they may be trapped in the clotting blood bolus or entangled in the developing peritrophic envelope, consisting of numerous peritrophic membranes. Later, degradation by proteases secreted into the midgut lumen is likely to occur, but this has not been tested. All *Culex* species seem to have the same time frame of peritrophic membrane formation (Houk et al. 1979). Adsorption to and invasion of the midgut epithelium of *Culex tarsalis* by St. Louis encephalitis virus seems to occur during the first 4 h after the ingestion of infected blood (Whitfield et al. 1973). The same seems to be possible with other adult mosquitoes and virus species, e.g. arboviruses. Hardy et

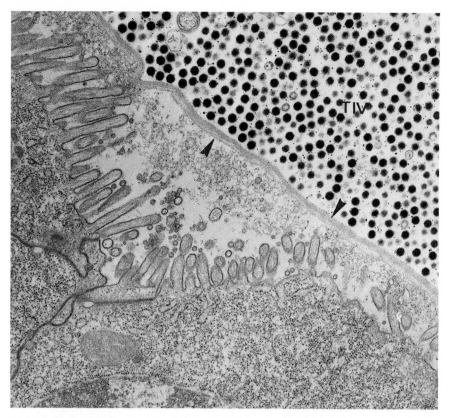

Fig. 74. The peritrophic membrane (*arrowheads*) of *Aedes taeniorhynchus* larvae acts as a barrier to fixed Tipula iridescent virus (*TIV*); 16 500x. (Electron micrograph courtesy of Dr. Stolz, Halifax, Canada)

al. (1978) reported genetically controlled resistance to infection of *Culex tarsalis* by western equine encephalomyelitis. The resistance factor was assumed to be related to the midgut.

6.6.2 Bacteria

Bacteria in the gut of insects may be pathogens, symbionts or just present. The former are of great interest for the control of insect pests or as pathogenic agents of useful insects (Brooks 1963; Steinhaus 1967).

Bacillus thuringiensis var. *entomocidus* is used as an insecticide as it synthesizes a parasporal crystal containing an endotoxin with cytolytic activities. After ingestion of these bacteria, the endotoxin destroys the midgut epithelium if it is able to penetrate the peritrophic envelope, consisting of many peritrophic membranes. Yunovitz et al. (1986) could show that in the larval midgut of *Spodoptera littoralis* (Lepidoptera, Noctuidae) larger aggregates of endotoxin proved to be unable to penetrate the peritrophic envelope. But a highly toxic fraction with M_r 64 kDa, which was obtained by using detergents, reached the midgut and destroyed it. The detergents used had no such effect. In vivo the small molecular fraction is obtained probably by proteolytic digestion.

The bacterial flora of the gut of the desert locust, *Schistocerca gregaria*, is restricted in the midgut to the endoperitrophic space and the lumen side of the peritrophic envelope (Hunt and Charnely 1981). Therefore, these authors suggested that the bacteria might not be able to penetrate the peritrophic envelope, or that the conditions in the ectoperitrophic space of the midgut were not suitable for them. In the hindgut a rich flora of Gram-negative rods and to a lesser extent of Gram-positive coccoids was present on the cuticle. This may be possible as the food bolus, as soon as it enters the hindgut, is broken up into smaller pellets by the action of the hindgut musculature.

A parasitic bacterium was described in the larvae of *Solenobia triquetrella* (Lepidoptera, Psychidae) by Puchta and Wille (1956). Infection occurred orally. The first instar was not infected. The midgut epithelium of the following stages seemed to be infected by bacteria which either circumvented the peritrophic membranes in the valvula region or in the anterior hindgut during ecdysis, when the old cuticle and epithelium were already detached. As the extrusion of these parts took a few hours, enough bacteria reached the new cuticle and epithelium and passed into the ectoperitrophic space. Therefore, first, cells of the most anterior and posterior parts of the midgut were invaded. Subsequently, the remaining cells of the midgut were infected. The endoperitrophic space, the regenerative cells and tissues other than midgut cells were free of bacteria. The peritrophic envelope itself could not be penetrated in this case.

An interesting approach was made in order to increase the susceptibility of gypsy moth larvae, *Lymantria dispar*, to infection with microorganisms in connection with pest control. In vivo and in vitro experiments were undertaken to degrade the peritrophic membranes of these larvae with chitinase of bacterial origin (Daoust and Gunner 1979). However, the alkalinity in the midgut is not compatible with the low pH required for the activity of chitinases. Therefore, an attempt was made to find bacteria which are able to lower the pH in the midgut of gypsy moth larvae. Bacteria isolated from the bacterial flora of the larval gut proved to be superior to various isolates from other sources. These were combined with chitinolytic bacteria for a synergistic microbial attack on gypsy moth larvae.

American foulbrood, caused by *Bacillus larvae*, is a lethal pathogen of larvae of the honeybee, *Apis mellifica*. Infection occurs by ingestion of spores, followed by germination of the spores in the center of the midgut lumen (Bamrick 1964). In first instars peritrophic membranes were observed after 36 h of age. If larvae were fed spores at the very early age of 3 h, bacteria were found to penetrate the peritrophic envelope and to invade the midgut epithelium after 60 h. The most anterior and posterior parts of the midgut epithelium were infected first. Heimpel (1955) also observed that the peritrophic envelope is penetrated especially at the posterior end of the midgut. After rapid colonization of the midgut epithelium, the bacteria invade the hemocoel mainly at metamorphosis, destroy adjacent tissues, and death of the brood follows. Young larvae are highly susceptible to *Bacillus larvae*, but older larvae are completely resistant. This is probably due to the extremely thick peritrophic envelope which consists of a very large number of membranes and which is tight at the posterior end (see also Sect. 3.3.2).

Another *Bacillus* species, *Bacillus sphaericus*, proved to be more or less toxic to mosquito larvae, *Culex quinquefasciatus* (*C. pipiens*). Strain SS II-1 killed at least 50% of these mosquito larvae in a laboratory assay using 10^3 bacteria/ml (Myers and Youston 1978). A toxin is responsible for the insecticidal activity. It is not released into the culture medium and is not present as a parasporal crystal (Singer 1973, 1974; Myers and Youston 1978). Davidson (1979) could show in an electron microscopic study that these bacteria, together with other bacteria which are harmless, are confined within the single, tube-like peritrophic membrane of these larvae. It was never observed that they appeared in the ectoperitrophic space. Within 30 min after ingestion by the mosquito larva the bacterial cell wall as well as the cytoplasmic ground substance disappeared when the bacteria were in the anterior and middle part of the midgut. Dadd (1975) demonstrated that in these mosquito larvae proteases are probably most active in the anterior midgut, whereas amylases are probably most active in the posterior part of the midgut. Therefore, Davidson (1979) assumed that the toxin of *Bacillus sphaericus* is released during digestion in the anterior midgut region. Nothing is known about the chemical properties or location and mode of action of the toxin. One hour after feeding on *B. sphaericus* most of these bacteria had been voided by defecation and only very few bacteria of this species were found in the midgut or hindgut. However, between 1 and 7 h after feeding, the total number of bacteria within the peritrophic membrane had increased markedly. Multiplication of *B. sphaericus* is obviously not necessary for pathogenesis. Between ingestion of these bacteria and the onset of symptoms of intoxication and histological changes, a lag phase of 7 to 10 h occurred. Around 10 to 12 h after ingestion of these bacteria, the larvae became sluggish, the midgut was swollen until it came to lie against the outer body wall; its cells were filled with cytolysosomes. Twenty-four hours after ingestion of *B. sphaericus* the larvae were either moribund or dead.

The effect of cold stress on the invasion of the hemolymph of houseflies, *Musca domestica*, by bacteria from the gut lumen was investigated by El-Mallakh (1984). Temperatures below 5 °C disturb the secretion process of peritrophic membranes of flies. The result is a "disturbed zone" of different extension (Zimmermann et al. 1973). This effect can be used for the determination of the formation rate of peritrophic membranes in flies and their larvae (Becker 1978a; see Sect. 3.3.6). El-Mallakh used *Salmonella typhimurium* as a marker because *Musca domestica* is an important carrier of this potential human pathogen. It is not pathogenic to the housefly and was fed from a broth culture. Cold exposures for different time periods and of different frequencies had no effect on hemolymph invasion by these bacteria. Five to 15% of the flies had bacteria in the hemolymph, and this percentage could not be influenced significantly by the variable cold treatment.

The marine wood borer, *Limnoria tripunctata* (Crustacea, Isopoda), is unique in its ability to feed on creosote-treated wooden structures (Zachary and Colwell 1979; Zachary et al. 1983). *L. tripunctata*, which feed on mangrove or other untreated wood, do not contain a gut microflora which is associated with the intestinal lining. An exception are those bacteria ingested with food which are enclosed by the peritrophic membrane. But 50% of the isopods from one sampling source which feed on creosote-treated wood had bacteria associated with the midgut epithelium.

Penetration of the peritrophic envelope, which was always intact, was never observed. Therefore, circumvention of the envelope has been assumed as a possible mechanism to invade the ectoperitrophic space. It seems possible that the acquisition and activity of these bacteria might contribute to the creosote resistance of the isopod.

6.6.3 Actinomycetes

An unexpected relationship between microorganisms and the midgut of a soil-feeding termite was reported by Bignell et al. (1980). The peritrophic envelope of *Cubitermes severus* from Nigeria is colonized on its epithelial face by actinomycetes. The peritrophic membranes act as an attachment site for the ramifying filaments of these organisms which extend through the ectoperitrophic space and form intimate associations with the microvilli of the midgut cells. The filaments were of different diameters, varying between 0.15 and 0.8 μm. They do not seem to be segmented and are rarely branched. The surface of the filaments is provided with pili which enable them to make contact with either peritrophic membranes or microvilli of midgut cells. These actinomycetes occurred in the midgut of all specimens of this termite. No pathogenic effects were observed. Therefore, the authors assumed that these organisms are either symbionts or commensals as the soil-feeding termites might be nutritionally dependent on symbiotic microorganisms. Nothing is known about the mode of infection, the way in which the ectoperitrophic space is reached nor the mechanism of adhesion to either peritrophic membranes or microvilli of the midgut cells (see Sect. 6.2).

6.6.4 Flagellata

After a blood meal infected with *Leishmania mexicana amazonensis* peritrophic membranes are secreted by the posterior midgut epithelium of the host, *Lutzomyia longipalpis*, which surround the blood completely (Killick-Kendrick et al. 1974). The ingested amastigotes transform into promastigotes which divide rapidly in the gradually digested blood. Some flagellates are embedded between peritrophic membranes, whereas in others only the flagellum is trapped between membranes. After about 3–4 days the blood is digested completely. The peritrophic envelope disintegrates and degenerated midgut cells are shed into the lumen of the posterior midgut. The microvilli of the midgut cells lengthen. The promastigotes leave the peritrophic membranes and attach to or insert between the long and slender microvilli. In this way the posterior midgut epithelium is colonized by the promastigotes after defecation of the remnants of blood, peritrophic membranes and degenerated midgut cells.

Nothing is known about the pathway of infection of *Herpetomonas ampelophilae* or other related Trypanosomatidae which live in the midgut of Diptera, either in the endo- or ectoperitrophic space or in the Malpighian tubules.

Trypanosomes of the *T. brucei* group were investigated for about 100 years, but nevertheless the highly complicated developmental pathways of these important pathogens have not been completely elucidated as yet. Evans and Ellis (1983) gave an excellent review of the history. During the era of light microscopy, the following version was accepted (Fig. 76):

After the ingestion of an infected blood meal by the tsetse fly some trypanosomes may survive in the crop for several days, but seem to be unable to establish themselves elsewhere. The majority is able to establish themselves in the midgut where they are separated from the midgut epithelium by the single, continuous, tube-like peritrophic membrane. They were believed to pass down the alimentary canal inside the peritrophic membrane and to circumvent the posterior end of the membrane in order to reach the ectoperitrophic space with its rich supply of low molecular nutrients. It was assumed that the trypanosomes pass forward in this compartment and multiply in it. Then they should penetrate the newly formed peritrophic membrane just where it left the cardia. The trypanosomes might then pass through the foregut into the hypopharynx and then by way of the ducts into the lumen of the salivary gland. In the gland transformation into the epimastigote and afterwards into the metacyclic trypomastigote form should take place. The infective metacyclics could leave the gland together with saliva. It was assumed that this complicated way is so dangerous that only a small percentage of trypanosomes could survive and hence only the observed, extremely small percentage of flies could be infective even in endemic areas.

It is astonishing that until 1929 the peritrophic membrane was never mentioned. In that year Wigglesworth described the formation of the peritrophic membrane in the tsetse fly and Johnson and Lloyd (see Evans and Ellis 1983) observed *T. congolense* in the ectoperitrophic space. Wigglesworth (1929) and Hoare (1931) believed that the peritrophic membrane is disrupted at its posterior end by the rectal spines.

However, several morphological and physiological observations in the past 30 years demanded a reexamination of the traditional pathway. It was found that newly emerged flies, fed within 4 h after emergence, could be infected to a higher percentage of up to 20% than older flies (Harley 1971). Willett (1966) could demonstrate that the formation of peritrophic membrane starts after hatching and that the posterior end of the newly formed membrane is completely closed. The formation rate is 1 mm/h during the first 30 h after emergence, and also after feeding. Therefore, most of a blood meal taken during that time must be pumped into the crop where the stumpy trypanosomes seem to have time to transform into a slender form and to change their metabolism from anaerobic metabolism, using mainly glucose, to aerobic metabolism, using proline as the main energy source. Electron microscopic investigations have shown that the trypanosomes are able to leave the endoperitrophic space by penetrating the peritrophic membrane (Fig. 75) in the anterior midgut region and most probably not in the cardia or adjacent to the cardia (Gordon 1957; Freeman 1973; Ellis and Evans 1977; Evans et al. 1979; Evans and Ellis 1983; Ellis and Maudlin 1985). Circumvention of the peritrophic membrane after it is disrupted at the posterior end by the spines of the hindgut is rather unlikely because there are adverse conditions for the trypanosomes in this region. Bursell and Berridge (1962) could show that the pH of the midgut contents is about 7.2 combined with low osmolarity, whereas in the hindgut a low pH of 5.8 and a very high osmolarity occurs. Furthermore, the highly complicated route of the trypanosomes to the salivary gland has been questioned during the last 20 years because trypa-

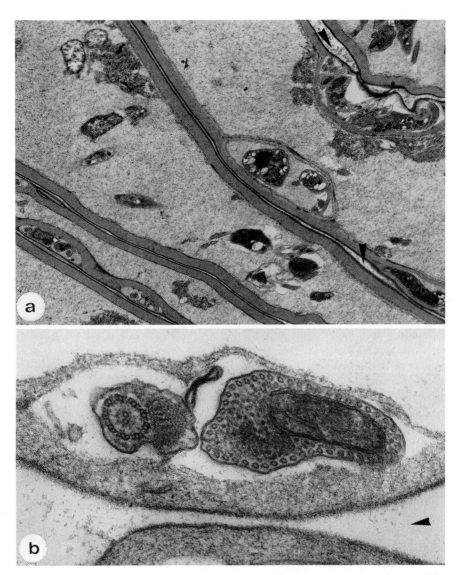

Fig. 75a,b. *Trypanosoma brucei rhodesiense* penetrating the peritrophic membrane of the tsetse fly, *Glossina morsitans morsitans.* The electron-dense layer represents the lumen side of the peritrophic membrane. The lumen, i.e., the endoperitrophic space, is marked by *arrowheads. a* 5200x; *b* 65 000x. (Electron micrographs courtesy of Dr. D.S. Ellis, London)

nosomes were found in the hemocoel of infected tsetse flies (Foster 1963, 1964; Mshelbwala 1972), and because tsetse flies could transmit trypanosomes after experimental infection of the hemocoel (Otieno et al. 1976). But the transition from midgut epithelium to the salivary gland might be dangerous, since Croft et al. (1982) found a trypanosomicide compound in the hemolymph of tsetse flies. Finally, it turned out recently that there are tsetse flies which are susceptible and others which

183

are refractory to infection with trypanosomes of the *T. brucei* group (Maudlin 1985). Selection and subsequent electron microscopy proved that those flies which are susceptible are infected with a rickettsia-like microorganism. Such microorganisms occur in the ovary and within cells of the midgut, particularly in the mycetome region. They are absent or present in low numbers in the ovaries of refractory flies. It remains to be elucidated how the metabolism of the rickettsia-like organism affects indirectly the trypanosomes. On the basis of these observations, Evans and Ellis (1983) proposed another pathway of infection (Fig. 76).

6.6.5 Sporozoa

Haemosporidia use mosquitoes as vectors and pass through gamogony in their midgut in a rather short time. The malaria parasites of the genus *Plasmodium* need about 24 h. The sister group, the Piroplasmida, has ticks as vectors. These parasites complete gamogony in the midgut of their vectors in about 10–14 days!

In mosquitoes the secretion of peritrophic membranes is triggered by an elevation of the salt concentration in the midgut lumen after a blood meal (Berner et al. 1983), whereas the secretion of proteases is triggered by the appearance of soluble proteins after a blood meal of the mosquito (Briegel and Lea 1975). Species-specific differences were found in the time needed for the completion of the peritrophic envelope and its persistence (see Sect. 3.3.2): in *Aedes aegypti* the liquid secretions solidify already 8 h after a blood meal; a discrete and uniform envelope is present about 12 h after the ingestion of blood (Perrone and Spielman 1986). It is degraded subsequently as blood digestion proceeds. *Plasmodium gallinaceum* has formed ookinetes about 14 h after the ingestion of infected blood (Stohler 1957). The malaria parasites seem to have to complete their gamogony before the peritrophic membranes surrounding the ingested blood harden, and before the concentration of proteases has reached a maximum in the midgut (Freyvogel 1980). Stohler found that ookinetes of *Plasmodium gallinaceum* are able to penetrate the newly formed peritrophic membranes of *Aedes aegypti*. No further passage was possible about 30 h after a blood meal. At that time many ookinetes were observed at and inside the peritrophic envelope. Probably a large number of ookinetes is unable to reach the midgut epithelium in order to undergo sporogony. This seemed to be the reason for the discrepancy between a high number of parasites in the infected blood and the considerably low amount of oocysts in the midgut.

In *Anopheles gambiae* a delicate envelope appears between 15–30 h after a blood meal and is degraded continuously afterwards. In *A. stephensi* a peritrophic envelope persists from 32–72 h after a blood meal and is degraded together with the remains of blood. It does not contain chitin but poly-N-acetylgalactosamine (Berner et al. 1983). In *A. atroparvus* the viscous peritrophic envelope does not harden. Sluiters et al. (1986) investigated the development of *Plasmodium berghei* in a refractory and a susceptible line of this mosquito species. These authors maintain that the incomplete and viscous peritrophic envelope cannot be considered to be a barrier to the ookinetes of *P. berghei*. Nevertheless, maximal 1% of the ookinetes developed into oocysts in the susceptible strain, but hardly any did so in the refractory strain.

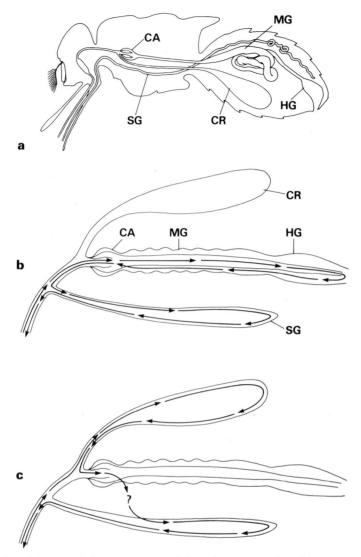

Fig. 76a-c. Infection of the tsetse fly by trypanosomes. Schematic representation of the tsetse fly and the main parts of its gut system. *CA* Cardia; *CR* crop; *HG* hindgut; *MG* midgut; *SG* salivary gland (After Geigy and Herbig 1955). *b* Traditional and *c* new hypothetic pathway. *b,c* After Evans and Ellis (1983)

Janse et al. (1985) have shown that the ookinetes of *Plasmodium berghei* develop rapidly in the midgut of their host. As the formation of the peritrophic envelope of its host, *Anopheles atroparvus*, is delayed, the peritrophic envelope is indeed no obstacle to these ookinetes. On the other hand, due to the late development and maturation of the ookinetes of *P. falciparum*, a large percentage of ookinetes is trapped in the peritrophic envelope of *Anopheles stephensi* (Meis and Ponnodurai

1987). Very few ookinetes were found in the ectoperitrophic space. The penetration of the midgut epithelium of *A. stephensi* and *A. gambiae* occurs intercellularly in this species, and therefore damage of the epithelium is prevented even with very heavy parasite loads. In *P. berghei* the intracellular penetration of the midgut epithelium of its host results in increased damage and mortality by increasing parasite loads (Maier et al. 1987). It would be interesting to investigate the development of ookinetes of other species of *Plasmodium* in susceptible and refractory strains of mosquitoes of the genera *Aedes* and *Culex*, as these have a peritrophic envelope which is mechanically more resistant than that of *Anopheles* species.

The investigations of Gass and Yeates (1979) and Yeates and Steiger (1981) have shown that proteases are able to destroy in vitro asexual stages and developing ookinetes. The results are in accordance with the in vivo situation in the midgut of mosquitoes. Immature ookinetes were found to be more sensitive to damage than mature ones which have developed already a typical sporozoan pellicle as described by Mehlhorn et al. (1980). The pellicle seems to protect the mature ookinetes for about 30 min. Therefore, it has been suggested that only ookinetes which developed in the middle of the blood clot, i.e. remote from the early peripheral protease activity, may have a chance to reach the midgut epithelium. However, it has still to be established if there are other factors involved in the drastic reduction of the population of parasites in the midgut of the mosquito.

The importance of the peritrophic envelope, if any, for the penetration of ookinetes of other Haemosporidia, of the genera *Haemoproteus* and *Leucocytozoon* has still to be settled Fallis and Bennett 1961; Desser and Wright 1968).

Piroplasmida use ticks as vectors. Rudzinska et al. (1982) could demonstrate electron microscopically that a peritrophic membrane is secreted by the midgut epithelium of larvae, nymphs and adult *Ixodes dammini* only after feeding. In larvae a first peritrophic membrane has a thickness of less than 100 nm. Fourteen to 16 h after repletion the peritrophic membrane of the larvae consists of several layers and reaches a total thickness of 200–300 nm. No fibrils could be detected. Cellular components, resulting from the breakdown of the ingested blood, do not occur in the ectoperitrophic space. Even ribosomes with a diameter of 15 nm do not pass through the peritrophic membrane, which therefore acts as an ultrafilter. Hemoglobin seems to permeate through the membrane. Microgamonts were obviously able to penetrate the peritrophic envelope with their arrowhead construction and were found afterwards in the ectoperitrophic space. The peritrophic membrane of this tick proved to be a transient structure which was absent 14–16 days after repletion.

6.6.6 Microsporidia

Nosema fumiferanae lives in the midgut of the eastern spruce budworm, *Choristoneura fumiferana*. Fourth to 6th instar larvae seem to be infected by the ingestion of frass-containing *Nosema* spores (Nolan and Clovis 1985). Where blebs of cytoplasma are extruded from infected cells, the peritrophic membranes are contaminated with microsporidia which were found embedded in the peritrophic envelope. Again, the peritrophic membranes do not function as a barrier against infection.

186

6.6.7 Larvae of Cestoda

Leger and Cavier (1970) tried to establish the oncosphaera of *Hymenolepis nana* in unusual hosts. Infection of *Gryllus domesticus*, *G. bimaculatus*, *Locusta migratoria* and *Schistocerca gregaria* by the ingestion of eggs resulted in the formation of cysticercoids which were larger than in the normal host, *Tenebrio molitor*. *Carausius morosus* could not be infected, neither per os nor by injection of oncospheres. Three species of cockroaches, *Periplaneta americana*, *Leucophaea maderae* and *Blaberus fuscus*, could not be infected per os but by the injection of larvae into the hemocoel. The authors assumed that in these cockroaches the peritrophic membranes might act as a barrier which could not be penetrated by the larvae in order to reach the midgut epithelium and hemocoel.

6.6.8 Microfilariae

Lewis (1950, 1953) was the first to show that blackflies, *Simulium damnosum*, secrete peritrophic membranes only after the ingestion of a blood meal. He suggested that these membranes might be responsible for the fact that normally only a small number of microfilariae of *Onchocerca volvulus* reach the flight muscles of the vector, which was confirmed by Bain and Philippon (1970) (Fig. 77a). Moreover, Bain et al. (1976) could show that the thickness of the peritrophic envelope secreted after a blood meal by *Simulium damnosum* is correlated with the number of microfilariae ingested. Uninfected blackflies had a very delicate and fragile peritrophic envelope, whereas in infected specimens its thickness increased considerably with the number of ingested microfilariae. In the savanna of West Africa only two microfilariae on average were found in the hemocoel of these blackflies, although up to 250 microfilariae might have been engorged during the blood meal. The authors assume that the activity of these microfilariae induces a considerable increase in the secretion of peritrophic membranes in this case and that the reinforced peritrophic envelope is responsible for the fact that only a small number of microfilariae reach the hemocoel of the vector (Fig. 76).

A special defense mechanism was found in *Simulium ochraceum* which is a vector of *Onchocerca volvulus* in Central America. Microfilariae are destructed by the action of cuticular teeth of the buccopharyngeal region (Omar and Garms 1975; Omar 1976).

In the meantime several authors investigated different species of vectors of microfilariae and parasites under different conditions. *Anopheles quadrimaculatus* needs only about 1 min for engorgement. The secretion of peritrophic membranes starts already soon after the ingestion of blood. Usually within 30 min, but not immediately after engorgement, a thick layer of viscous material is secreted which separates blood bolus and midgut epithelium. But after a few hours it becomes thinner, measuring about 60 μm in thickness, and results in a tough, multilayered peritrophic envelope which remains intact for at least 24 h (see also Sect. 3.3.2). Most of the microfilariae of *Brugia pahangi* leave the blood mass already 5–15 min after ingestion of an infected blood meal. They penetrate the posterior part of the midgut and use a cephalic hook for the penetration of the thin, viscous peritrophic

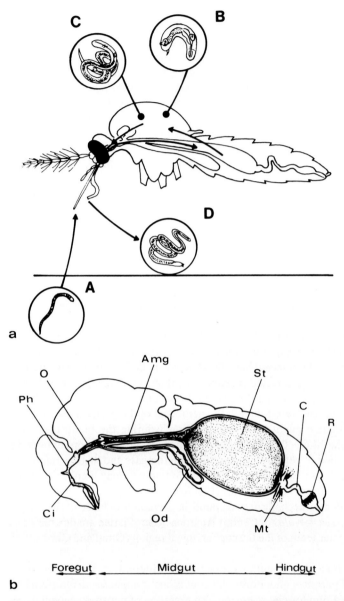

Fig. 77a. Transmission of filariae by mosquitoes. *A* Microfilariae in the blood of a vertebrate host; *B* and *C* first and second larval stages in the flight muscles of the vector, *D* the third larval stage is transmitted to a new vertebrate host when the mosquito bites (after Bain 1986). *b* Diagrammatic representation of the alimentary tract of a blackfly, *Simulium metallicum*. *Amg* Anterior (tubular) midgut; *Ci* cibarium; *C* colon; *O* esophagus; *Od* esophageal diverticulum; *Mt* Malpighian tubules; *Ph* pharynx; *R* rectum; *St* stomach (expanded midgut) (After Omar and Garms 1977)

envelope which is still in statu nascendi, and of the midgut epithelium (Esslinger 1962). Such hooks were also found in the microfilariae of other species (*Brugia malayi, Wuchereria bancrofti, Loa loa, Dirofilaria immitis* and *Litomosoides carinii*). They cause heavy destructions of the peritrophic envelope, the midgut epithelium and the basement membrane after mass infection. Thirty minutes after a blood meal only very few microfilariae remain in the midgut.

Christensen and Sutherland (1984) found that about 50% of the microfilariae of *Brugia pahangi* ingested by *Aedes aegypti* left the midgut within 2.5 h after an infected blood meal and migrated into the hemocoel and flight muscles. But in *Aedes trivittatus* infected with microfilariae of the same species only 7.5% of the ingested parasites reached the hemocoel (Sutherland et al. 1986). Light microscopy revealed only few quantifiable differences in peritrophic membrane formation between the two mosquito species. If the mosquitoes were fed on heparinized or citrated blood, a significant increase in the percentage of microfilariae penetrating the midgut of *A. aegypti* resulted, whereas in *A. trivittatus* a significantly reduced percentage of microfilariae was able to penetrate the midgut. After feeding blood-containing EDTA the percentage of microfilariae which were able to penetrate the midgut was reduced significantly in both species of mosquitoes. Therefore, the peritrophic envelope seems to be only an inferior barrier to the microfilariae on their way from the midgut lumen to the hemocoel and flight muscles of the mosquito.

The sheath which covers microfilariae of *Brugia pahangi* and *B. malayi* seems to remain in place until the microfilariae have penetrated the midgut epithelium and reached the hemocoel (Yamamoto et al. 1983; Christensen and Sutherland 1984; Agudelo-Silva and Spielman 1985). Exsheathment might occur during penetration of the gut when the sheath is caught and held by the surrounding tissue and the microfilaria moves anteriorly until it ruptures the anterior end of the sheath and is able to leave it eventually (Perrone and Spielman 1986).

In order to determine the postulated effect of the peritrophic envelope on penetration of the midgut by microfilariae, Perrone and Spielman (1986) fed ferritin as an electron-dense marker together with the blood. Fifteen to 20 min after the blood meal ferritin particles could be observed evenly distributed electron micros-copically between the ingested, tightly packed erythrocytes and the surface of the midgut epithelium. Microfilariae were found in the clear zone filled with ferritin, but no traces of peritrophic membranes appeared during this early phase of infection. Therefore, in this case peritrophic membranes cannot be a barrier against the invasion by microfilariae.

In *Anopheles stephensi* 80–100% of ingested microfilariae from reptiles, *Foleyella furcata*, reached the hemocoel within 3 h after ingestion (Bain and Philippon 1970).

On the other hand, increased mortality of the vectors is correlated with the number of ingested microfilariae (reviews by Maier 1976; Omar and Garms 1977). *Aedes trivittatus* showed considerable tolerance to infection with microfilariae of *Dirofilaria immitis* (Christensen 1978). A reduction of the parasite burden occurred by the peritrophic envelope and resulted in increasing longevity of the mosquitoes. However, mortality of the vector increased with the number of microfilariae ingested with infected blood from dogs.

The rapid death of blackflies after heavy infections may be due to the following reasons: (1) massive invasion by microfilariae may cause serious injuries or even the rupture and destruction of the midgut epithelium. (2) Interruption of the formation of the peritrophic envelope, particularly at the anterior or posterior end, blood flow into the hindgut and the disruption of the hindgut. (3) Invasion and destruction of various organs of the blackfly by excessive numbers of microfilariae (Omar and Garms 1977) (Fig. 77b).

The peritrophic envelope does not seem to be an obstacle for most of the viable microfilariae which use the first minutes after infection to penetrate from the lumen of the midgut into the hemocoel. *Onchocerca volvulus* and its vector *Simulium damnosum* seem to be an exception.

6.6.9 Summary

Parasites which develop and multiply in the midgut epithelium or in the hemolymph must penetrate the peritrophic membranes. This seems to be easy in the case of peritrophic envelopes consisting of loosely packed peritrophic membranes secreted in the form of patches by the whole midgut epithelium. It is also possible during ecdysis and in the case of blood-sucking mosquitoes in a few hours after a blood meal when the secretion of material for the peritrophic membranes has already started but not yet resulted in the completion of these membranes. The continuous tube-like peritrophic membranes of the Dipteran type were believed to be impermeable to parasites. It was assumed that parasites had either to circumvent the membrane in the hindgut or to use holes in these membranes, regardless of the origin of such holes. However, such mechanisms do not seem to exist. Some pathogenic microorganisms seem to be able to penetrate these membranes and larger parasites were observed even to rupture peritrophic membranes or envelopes.

But it should be noted that even partial protection from parasites would help to reduce the danger of a massive infection which could hamper or even kill the insect vector, for instance by the destruction of its flight muscles.

The coordination of the time frame of formation of peritrophic membranes and penetration of parasites is of great importance. Those parasites which have no development in the midgut, e.g. microfilariae, try to penetrate the midgut as soon as they arrive in this region of the gut. However, parasites like Haemosporidia, which have to pass through gamogony, must reach the stage of a mature ookinete with a complete pellicle as soon as possible, in about 24–48 h, and then penetrate the peritrophic envelope and the midgut epithelium. The pellicle is needed to withstand at least for a short time the action of proteases. The Piroplasmida, probably the sister group of the Haemosporidia (Mehlhorn et al. 1980), need much more time to complete their gamogony than the Haemosporidia, about 10–14 days. This is perhaps one of the reasons why they use ticks instead of mosquitoes as vectors.

As a result of probably long-time selection, the pathogenic parasites are able to penetrate the peritrophic membranes. However, no one has regarded related species which are either poor vectors or unable to transmit such parasites. For instance, it is still unknown why tsetse flies or blood-sucking bugs do not transmit

190

microfilariae (Lavoipierre 1958) or Piroplasmida. It is remarkable that in some vectors, such as lice and adult fleas, peritrophic membranes are lacking.

6.7 Fecal Pellets or Ecological Aspects of Peritrophic Membranes

At first sight fecal pellets appear to be useless remnants of digestion. However, in the past 20 years more and more evidence accumulated that fecal pellets of zooplankton play a considerable role in aquatic systems, in marine as well as freshwater systems. Much less is known about their role in terrestrial systems.

6.7.1 Aquatic Ecosystems

1. Fecal pellets may still contain considerable amounts of useful nutrients.
2. The residual nutrients in fecal pellets are converted by bacteria and other microorganisms mainly from carbohydrate to more nitrogen-containing compounds.
3. Fecal pellets may be ingested and digested by other members of the zooplankton.
4. Fecal pellets are responsible for a considerable acceleration of the sinking rate of their ingredients and reach the bottom of even deep freshwater lakes or oceans where they are valuable food for benthic detritus feeders.
5. Fecal pellets contribute to a depletion in surface water of certain elements, such as phosphorus and silica, which are recycled by the upwelling of dissolved material after some time.
6. Furthermore, fecal pellets are also involved in the transfer of pollutants.

Therefore, fecal pellets are no longer regarded as unimportant waste but as a significant source of food in aquatic ecosystems (see the excellent review given by Turner and Ferrante 1979). They may be included in larger aggregates called marine snow. These range from 1–4 mm in diameter and consist of phytoplankton, detritus, bacteria, and fecal pellets embedded in mucous material.

6.7.2 Fecal Pellet Membranes = Peritrophic Envelopes

Food residues are generally enveloped in the midgut by peritrophic membranes or mucus and dehydrated while transported through the hindgut. The number of peritrophic membranes in a given peritrophic envelope, i.e. the "membrane" of the fecal pellet, is quite different. It can be determined only by electron microscopy. The food residues are voided either as strands or as fecal pellets which are formed by peristaltic movements of the midgut or hindgut. Therefore, it is possible to obtain such material not only by collecting egested fecal pellets but also by preparation from freshly caught or even preserved specimens (Moore 1932). Usually, fecal pellets range in length from 50 to 3000 μm (Moore 1931b). The pellets may have species- or stage-specific size and form. It is possible sometimes to discern the origin of fecal pellets which are found in marine or freshwater sediments. Moore (1932) reported that this was noted already in 1678 by Martin Lister in his *Historiae Animalium Tractatus* in the case of certain mollusks. Moore (1931b,c) gave descriptions of the morphology of the fecal pellets of more common British marine Mollusca. These show a wide range of form, from the simplest type of rod-like to ribbon-like or bead-like pellets, and rods sculptured with complex transverse, longitudinal or spiral patterns. Fecal pellets of Trochidae are characterized by the internal separation of coarse and fine material in different parts of the pellet. Bandel (1974) investigated the different forms of fecal pellets of 109 species of Mollusca (Amphineura and Prosobranchia) from the Caribbean Coast of Columbia. Amphineura produce plain ovoid and ellipsoid pellets, Archaeogastropoda plain or sculptured rods, Mesogastropoda mainly plain and sculptured ovoids and spindles, and Neogastropoda — with very few exceptions — soft, mucoid, feces. The type of food and the kind of feeding and the ecology were considered in relation to pellet form and consistence. Thus, Schäfer's (1953) assumption that all sedentary or slow-moving gastropods form hard, solid pellets or mucoid follicles in order to avoid contamination of their own environment with fecal waste material could be disproved by Bandel for the Caribbean. Schäfer investigated the North Sea fauna. Under certain conditions only one or very few pellet types may occur. This has been reported for a lagoon on the Bahamas by Kornicker and Purdy (1957), where up to 97% of the sediment consisted of fecal pellets of *Batillaria minima*.

In Mollusca some relationships between taxonomy and shape of fecal pellets have been demonstrated (see Moore 1931b,c; Bandel 1974), especially in Amphineura and Archaeogastropoda. Abbott (1954) used fecal pellet characters together with other features to remove the gastropod genus *Echinius* from the family Modulidae to place it in the Littorinidae. Although fecal pellets may be of phylogenetic importance (Kornicker 1962), the restricted amount of knowledge does not allow further conclusions.

Among most of the Crustacea the fecal pellets are rather simple rod-like forms which may be tapered at the ends: cirripeds like *Balanus* spp., mysids and most species of the Decapoda, except some Anomura (Moore 1932). Fecal pellets may attain 1 cm in diameter, for instance in a crab from Borneo, *Thalassina anomala*. Martens (1976) described the fecal pellets of four dominating species of Copepoda of the Bay of Kiel: *Acartia bifilosa*, *Centropages hamatus*, *Oithona similis* and *Pseudocalanus elongatus*. Among copepods, the pellets are either rod-like, sometimes with a rounded end or drawn out at the end to a thin point. In Lake Michigan the most abundant calanoid copepod of surface waters throughout the year was *Diaptomus sicilus* which produced fecal pellets with a length ranging between 80 to 160 μm (Ferrante and Parker 1977). These could not be confounded with cyclopoid fecal pellets which proved to be very delicate and were destroyed by slight agitation in a towed plankton net. In deep waters of the same lake *Limnocalanus macrurus* was the most abundant species (75–100%) which voided large fecal pellets (150–200 μm long). Fecal material of salps is extruded from their gut as a continuous flattened ribbon. As this breaks at thinner portions, fecal pellets are formed which vary in length but are constant in width and thickness in the respective stage and species (Madin 1982). In the genera *Salpa* and *Pegea* the material is better packed than in *Cyclosalpa* which produces loose and flocculent fecal matter. In *Salpa maxima* veritable bricks are formed. Salps occur ubiquitously in the oceans. Normally, their density is fairly low, but periodically dense swarms occur in continental shelf and slope areas (Wiebe et al. 1979). Recent quantitative investigations have shown that salps which were neglected previously, in fact, play an important role in the overall vertical transport of material from the surface region to the seafloor due to the high sinking rates of their fecal pellets (Silver and Bruland 1981). Dunbar and Berger (1981, cited in Madin 1982) found in sediment traps at a depth of 341 m in the Santa Barbara Basin, California, that 60% of the fecal pellets were from salps, 25% probably from copepods and 15% of unknown origin, and that the fecal pellets from salps resulted in an organic carbon flux of 60 mg m^{-2} day^{-1}. Salp fecal pellets were also the main constituent in sediment traps at 200-m depth at a station in the North Pacific and constituted a carbon flux of 10.5 mg m^{-2} day^{-1} (Iseki 1981). Swimming salps filter water almost continuously through the mucous filter in their gut. In the laboratory Harbison and McAlister (1979) could show that *Cyclosalpa* spp. are able to retain particles larger than 4 µm with 100% efficiency. This fine mesh filter is able to retain small coccolithophorids, such as *Emiliani huxleyi*, with a diameter less than 6 µm, and even smaller particles (Silver and Bruland 1981). However, these authors did not consider that in Tunicata two types of envelope are formed in the gut. Food is wrapped into the mucous sheets in the foregut and these are surrounded in the midgut by several peritrophic membranes with a texture of chitin-containing microfibrils embedded in a protein-containing matrix. The latter make up the peritrophic envelope or membrane of the fecal pellet. Pteropods, such as *Corolla spectabilis*, are gastropods which build large, unsupported external webs. These collect particles passively. The effective pore size of such webs, which is in the range of 10 µm, does not allow the filtration of coccolithophorids, but mainly larger phytoplankton, especially diatoms (Silver and Bruland 1981). Therefore, in the same area, the central California Current, vertical transport by fecal pellets results in a greater calcite flux if salps are abundant, and a greater silica flux if pteropods are abundant.

6.7.3 Production

The number of fecal pellets produced per day not only depends on the species and size but also on the type of food. Reeve (1963) found that in *Artemia salina* large stages produced 10 fecal pellets per day if *Dunaliella* was fed at a particle concentration of 500 per ml, that 20 pellets were voided per day if *Chlorella* was fed at the same concentration, and 100 pellets per day were produced if only sand particles were fed. The number of pellets voided per day was roughly the same if algae were fed in a concentration range between 10 and 1000 particles per ml. But if sand was fed, the production of pellets increased from 20 to 100 in the same concentration range. The nature of the ingested food particles had no effect on the shape and size of the fecal pellets. It is interesting that over 3 weeks the number of fecal pellets voided was roughly the same in fed and unfed animals.

Marshall and Orr (1972) reported a maximum production of 12 pellets per hour per copepod, *Calanus finmarchicus*.

Georgi (1969b) observed that unfed crayfish, *Cambarus affinis*, voided empty fecal pellets throughout winter if held at low water temperature.

Frankenberg et al. (1967) could show that fecal pellets not only provide a significant source of food in plankton systems but also in species living in the sand of beaches. The authors investigated the production of fecal pellets by a burrowing shrimp, *Calianassa major*. This species occurs sometimes in great abundance along the southeastern coast of the USA. The entire burrow has a length of about 140 cm and usually a small and inconspicuous opening which is indicated by a ring of fecal pellets. Approximately 456 ± 118 pellets were produced per burrow and day. As the receding tide washed away most pellets from the beach into the subtidal zone, it has been estimated that at least 70% of the total feces produced or 7.84 g of organic carbon per day is exported into the subtidal zone.

6.7.4 Contents of Fecal Pellets

The identifiable contents of fecal pellets are nearly always undigestable inorganic skeletons such as diatom shells or coccoliths, whereas the remains of the main components of the original food of copepods and other crustaceans, flagellates and ciliates cannot be identified as they are digested almost completely (Bathman et al. 1987). Nevertheless, analysis of only the undigestable and definable fecal pellet contents, although fragmented by the mouthparts prior to ingestion, contributed valuable data concerning the question of food selection by copepods. It turned out that marine (Turner 1986) as well as freshwater copepods (Ferrante and Parker 1977) may be nonselective suspension feeders. Turner, in a series of papers, examined the feeding ecology of several copepod species illustrated by excellent scanning electron micrographs. For instance, the fecal pellets of *Undinula vulgaris*, a copepod from the continental shelf waters of the Gulf of Mexico, contained a broad size and shape range and a taxonomic array of phytoplankton (Turner 1986). The fecal pellets of calanoid copepods from Lake Michigan (mainly *Diaptomus sicilus*) contained 21 of the known 30 species of diatoms in this lake (Ferrante and Parker 1977).

Usually, food remains in the gut for longer periods if food is only scarcely available or in starved animals. In *Artemia salina* colored food could be observed in the gut even after 2 weeks of starvation, although under such circumstances the gut contents appeared to be highly diluted (Reeve 1963).

Honjo (1978) found two different types of fecal pellets in deep-sea sediment traps. The "green pellets" contained fresh diatom and coccolith remains which might have been voided by zooplankton grazing in the surface layers. "Red pellets" were found only in depths below 2800 m and contained mainly clay minerals and only little plant material. Honjo assumed that the red pellets were egested by coprophages living in deep water and that these had previously ingested the remains of former green pellets. Bioluminescence of bacterial origin was detected in 70% of all samples of freshly excreted fecal pellets from marine copepods and in materials collected from sediment traps (Andrews et al. 1984).

6.7.5 Pollutants in Fecal Pellets

Coastal waters and lakes are threatened more and more by many types of pollutants. Certain amounts of these are incorporated by organisms and accumulate in certain parts of the food webs (Vernberg and Vernberg 1974). As fecal pellets proved to serve in the transport of organic matter from surface waters into deeper layers, they might be involved in the transport of pollutants into deep-sea organisms far from the source of pollution (Turner and Ferrante 1979). It has been shown that fecal pellets may contain elevated concentrations of heavy metals (Boothe and Knauer 1972; Small and Fowler 1973; Benayoun et al. 1974; Fowler 1977), radionuclides (Osterberg et al. 1963; Keckes et al. 1972; Fowler et al. 1973; Benayoun et al. 1974; Cherry et al. 1975; Heyraud et al. 1976; Higgo et al. 1977; Beasley et al. 1978), chlorinated hydrocarbons (Baird et al. 1975; Elder and Fowler 1977) and even petroleum hydrocarbons (Conover 1971, cited in Turner and Ferrante 1979). Conover assumed that about 20% of the spilled oil from the wreck of a tanker might be incorporated into fecal pellets of copepods and transferred to the bottom of the sea.

Sometimes the concentrations of these pollutants exceeded those in animals which had formed these pellets. Boothe and Knauer (1972) investigated the accumulation of heavy metals in kelp crabs, *Pugettia producta*, which had been fed on the brown alga *Macrocystis pyrifera*. Except for Cr and Cd, whose concentrations in the fecal pellets was less than in the kelp, all other elements (Co, As, Zn, Cu, Pb, Fe) showed levels in the pellets 2.3–5.4 times those in the algae. Only for Fe the level in the pellets was 14 times that in the algae.

More than 90% of the zinc flux (Small and Fowler 1973) and 84% of the cadmium flux (Benayoun et al. 1974) through the euphausiid *Meganyctiphanes norvegica* was accounted for by fecal pellet deposition on the seafloor in the Ligurian Sea (Mediterranean Sea).

The naturally occurring isotope ^{210}Polonium is present in seawater at relatively low levels. This alpha-emitting isotope is accumulated by most marine organisms. Concentration factors of 10^3 to 10^4 are common. Fecal pellets constitute the main elimination route and play an important role in transporting ^{210}Po from surface layers to the seafloor (Heyraud et al. 1976).

Turner and Ferrante (1979) suggested that pollutants accumulated in zooplankton fecal pellets might be transported downward in the water column as well as upward through food chains. They also pointed out that these possible ways of transportation should be borne in mind where benthic organisms such as shrimps, lobsters, bivalves and fish are used as human food.

6.7.6 Sinking Rates

The measurement of particle fluxes is of greatest importance in modern marine biology. Therefore, a great number of investigations were carried out to determine not only sinking rates but also the effects of hydrodynamic as well as biological factors which influence the measurement and interpretation of sedimentation data.

It has long been a matter of speculation how deep-sea organisms are able to obtain food in the complete darkness of their environment. It was assumed that only predators might survive under these conditions. More recently, it could be demonstrated that many species of inshore marine teleosts and larger benthic invertebrates are coprophages. Surface waters show an abundance of organisms due to the use of sunlight for photosynthesis by algae and a multitude of food chains based on the products of photosynthesis. Only since about 1970 more detailed investigations were carried out on the deposits as well as chemosynthetic microorganisms at the bottom of oceans and deep lakes. These showed that deep-sea animals do not depend on the remains of carcasses which sink from surface waters into the darkness. It now becomes clear that a constant rain of the remains of zooplankton food sinks to the abyssal depth which not only allows the existence of benthic forms but also of other animals and microorganisms in the water column between surface and ground. The question is how sinking organic particles which are subject to microbial degradation, mechanical disruption by turbulence, and flocculation, can reach abyssal depths. Only fast-sinking particles should be able to reach deeper regions with appreciable amounts of organic matter usable as a source of food for deep-sea organisms. Therefore, a rapid sinking mechanism had to be postulated. Obviously, fecal pellets originating from zooplankton provide such a mechanism. The remains of food, organic and inorganic matter, are packaged into fecal pellets and can thus be translocated with a sufficient sinking rate to the deeper parts of the water column and at last to the ground region of oceans and deeper lakes. Mechanical fragmentation by the mouthparts of planktonic Crustacea meets at least partly the requirements for the recycling of their contents by microorganisms.

Phytoplankton cells sink at rates of less than one to only tens of meters per day (Smayda 1970, 1974), whereas fecal pellets of copepods may sink at rates ranging from 5–220 m/day (Osterberg et al. 1963; Smayda 1969, 1971; Fowler and Small 1972; Wiebe et al. 1976; Ferrante and Parker 1977; Turner 1977; Honjo and Roman 1978; Small et al. 1979; Bienfang 1980; Krause 1981; Bathelt and Schelske 1983). In freshwater copepods from Lake Michigan, *Diaptomus sicilus*, the sinking rate of fecal pellets collected in sediment traps ranged between 4.3 and 5 m/day (Ferrante and Parker 1977). Bathelt and Schelske (1983) determined a sinking rate for fecal pellets of various copepods from the same lake ranging from 27 to 82 m/day.

The large fecal pellets of salps which may have a length between 1 and 13 mm and a width between 0.3 and 4.5 mm sink much faster than the smaller ones from crustaceans. Bruland and Silver (1981) and Madin (1982) determined the sinking rates of fecal pellets from several species of Pacific salps which varies between 320 to 2700 m/day. Fecal pellets of pteropods had a sinking rate of 1800 m/day and those of doliolids 208 m/day.

Recoveries of substantial numbers of fecal pellets in deep-sea sediment traps, water samples or material from benthic habitats down to 5000 m showed clearly that fecal pellets reach the abyssal parts of the oceans (Schrader 1971; Wiebe et al. 1976; Soutar et al. 1977; Bishop et al 1977, 1978; Honjo 1978; Knauer et al. 1979). The origin of such pellets from surface layers could be demonstrated by the remains of phytoplankton in fecal pellets collected at 450 m in the Baltic Sea and at 4000 m in the Atlantic off Portugal (Schrader 1971), at 2000 m near the Bahamas (Wiebe et al. 1976) and at 5367 m in the Sargasso Sea (Honjo 1978). Recent studies have shown that sedimentation of fecal pellets is of great importance in the flux of both organic and inorganic matter to the deep sea and ocean floor (Hall et al. 1977; Bishop et al.

1977, 1978; Honjo 1978; Turner and Ferrante 1979). Honjo estimated in 1977 that 92% of the coccoliths produced in the euphotic layer of the equatorial Pacific are transported to the underlying seafloor in rapidly sinking fecal pellets, but revised his opinion in 1987 as more recent data suggest that this role is minimal (see Pilskaln and Honjo 1987). The probability that fecal pellets reach the seafloor depends on three independant parameters: the sinking rate, the rate of degradation, and the extent of coprophagy (Bathmann and Liebezeit 1986). A review of fecal pellet sinking rates from various geographical locations was given by Angel (1984). Of course, the sinking rate depends on size, form, contents and the specific weight of the respective fecal pellets but also on hydrographic conditions. The specific weight is relatively large in fecal pellets which are filled with diatom frustules. The sinking rate in undisturbed water of fecal pellets containing diatoms had a considerable range between 15–153 m/day (Turner 1977). Moreover, it could be shown that the sinking rate is enhanced by the cylindrical, tapered shape of fecal pellets which reduces the frictional drag. Several authors have shown that fecal pellets from copepods do not always sink from the surface zone (Smetacek 1980; Krause 1981; Bathmann and Liebezeit 1986). This may be due to coprophagy by other copepods or salps (Silver and Bruland 1981) or due to the rapid degradation by microorganisms and following disintegration. On the other hand, vertical flux occurs for some fecal forms, for instance the fecal strings of Euphausiids (von Bodungen 1986). Fecal pellets do not sink rapidly if they are very small, so-called minipellets (Gowing and Silver 1985), or if gas production occurs inside the pellet during microbial degradation which increases the buoyancy (Krause 1981). However, the latter assumption seems to be unlikely as there should be rapid diffusion and no accumulation of gases in fecal pellets (Lampitt et al. 1990). Alldredge et al. (1987) reported evidence for sustained residence of large macrocrustacean (probably euphausiid) fecal pellets in the surface waters off southern California. These fecal pellets were observed at abundances ranging from 500 to 98 000 pellets m^{-3}. They were produced by vertically migrating macrocrustaceans at night.

Although small particles dominate the standing stock of material in the sea, it is the large particles which are responsible for the majority of the vertical flux of material (Lampitt et al. 1990). The large fecal pellets of salps are fairly flat rectangles which sink faster than any pellets from other zooplankton members (see above). But they sink more slowly than would be expected from their size, in fact, two- to fourfold slower than predicted by the Stokes equation for a sphere of equivalent size and density (Dunbar and Berger 1981, cited in Madin 1982). This is not only be related to their shape but also to their specific weight. The density of freshly produced fecal pellets from *Salpa fusiformis* was determined by Bruland and Silver (1981) to be about $1.10 g cm^{-3}$. Dunbar and Berger determined the density of fecal pellet material from salps obtained from sediment traps to be about $1.23 g cm^{-3}$, about the same as for copepod fecal pellets (Komar et al. 1981).

In times of a phytoplankton bloom or due to an upwelling system, food passes the gut of zooplankton almost unchanged and is packaged into fecal pellets of high nutritive value. Beklemishev (1962, cited in Turner and Ferrante 1979) suggested already that this type of superfluous feeding could result in an improved food supply for mesopelagic and abyssal organisms. Diurnal vertical migration of zooplankton might contribute to this transport of organic matter into deeper water (Turner and

Ferrante 1979). However, superfluous feeding is a matter of controversy since assimilation of food substances by copepods did not decrease even at high concentrations of food (see Turner and Ferrante 1979).

6.7.7 Digestion-Resistant Components

The viability of marine phytoplankton in fecal pellets was investigated by Fowler and Fisher (1983). Fecal pellets often contain numerous intact cells, eukaryotic algae, Gram-negative bacteria and unicellular blue-green algae (see Silver and Bruland 1981). Electron microscopy showed that these cells appeared to be unaffected by digestion. Their cell walls seem to be able to exclude digestive enzymes. In fact, the cellulose plates of thecate Dinoflagellata are underlain by layers of sporopollenin, a material which is highly resistant to digestive enzymes. If such intact algae and bacteria are packaged into fecal pellets and transported into deeper layers, a considerable vertical flux of biomass results. Silver and Bruland estimated that an input of at least several mg carbon m^{-2} day^{-1} of intact cells occurred in deep waters. The benefits of digestion resistance are unknown. It may have survival value in upper layers, but if such cells are transferred to deep-sea regions no survival appears possible.

6.7.8 Nutritional Content

Concurrent with the amount of phytoplankton, the amount and organic content of fecal pellets voided by zooplankton change. In Bedford Basin, Nova Scotia, fecal pellets were the predominant material deposited in sediment traps during summer (Hargrave and Taguchi 1978). Due to the high organic content of these pellets oxygen consumption was high (Hargrave 1978). During a red tide in Tokyo Bay, copepods produced fecal pellets of high organic content. The microbial decomposition caused an oxygen-free bottom layer (Seki et al. 1974).

The nutritional content of fecal pellets varies considerably. Depending on the amount of phytoplankton, food is utilized more or less thoroughly. Consequently, the C/N ratio of fecal pellets is a good measure to estimate the nutritive value as well as the metabolic history (Knauer et al. 1979). If phytoplankton is abundant, zooplankton fecal pellets contain a high amount of undigested material such as proteins, carbohydrates and chlorophylls. Therefore, the C/N ratios approach the ratio for living plankton 106:16 = 6.6. When food is scarce, degradation and assimilation take place more thoroughly in the gut of zooplankton members. Nitrogen-containing compounds are extracted more completely and the C/N ratio increases. Knauer et al. (1979) found that the C/N ratios of deposited materials ranged from 11 to 30 and that the C/N ratio increased progressively with depth. In suspended particulates a similar increase with depth in the C/N ratios was observed by Bishop et al. (1977, 1978). C/N ratios of fecal pellets from salps of the Gulf stream in the North Equatorial Atlantic fell in a range of 10 to 12.5 (Madin 1982). Protein (about 60%) and carbohydrate (about 30%) were the major biochemical constituents of these pellets.

After a spring bloom phase in the Baltic Sea, consisting mainly of various species of diatoms, a significant portion of ingested food was voided with fecal pellets by the copepod *Pseudocalanus elongatus minutus* without being assimilated. A high chlorophyll a content in the pellets could be related to the high number of intact diatoms present (Bathmann and Liebezeit 1986). These intact diatoms could also be demonstrated in scanning electron micrographs. About 2 weeks later phaeophorbide, the product of chlorophyll degradation by zooplankton, dominated in the pellets.

The carbon content of particulate matter collected in sediment traps was measured by Wiebe et al. (1976). It was estimated that 7–25% of the carbon reaching the seafloor was of organic origin and that fecal pellets contributed a considerable percentage of the total energy requirements of the benthos. Bathmann et al. (1987) used a permanent station in the Norwegian Sea during May and June 1986. During May, 10 000 fecal pellets/m^3 and day were counted and 15 mg fecal pellet carbon (FPC)/m^3 and day was measured in the surface layer. These values decreased in lower layers: at a depth of 100 m 2000 fecal pellets/m^3 and day or 2 mg FPC/m^3 and day and in depths of more than 500 m these values declined to almost zero. During June, a decrease was observed to less than 1 mg FPC/m^3 and day near the surface.

6.7.9 Colonization of Fecal Pellets by Microorganisms

The colonization of fecal pellets by microorganisms is of greatest importance with respect to the nutritional value and the C/N ratio (Ferrante and Parker 1977; Honjo and Roman 1978; Turner 1979). It has been demonstrated by many authors using scanning electron microscopy but also by using tritium-labeled adenine as a marker (Andrews et al. 1984) that fecal pellets become colonized already soon after egestion. The latter authors found that bacterial activity declined rapidly during the first 10 h after egestion. Freshly egested pellets from copepods proved to be free of bacteria (Ferrante and Parker 1977; Lautenschlager et al. 1978; Honjo and Roman 1978; Turner 1979), whereas pellets from the shrimp *Palaemonetes pugio* were densely packed with bacteria immediately after voiding (Johannes and Satomi 1966). Depending on the temperature in the environment, the degradation of the peritrophic envelope of copepods is said to take place in a few hours or several days if surface waters of lakes or the sea are in the range of 20 to 25 °C. In deeper layers with a temperature of about 5 °C, the pellet envelope may remain intact for about 20 to 35 days (Honjo and Roman 1978; Turner 1979). The degradation of the peritrophic envelope, which takes place during the first hours or even days after microbial colonization, results in a significant reduction in carbon content accompanied by minor changes in nitrogen content. This could explain C/N ratios of less than 6 (Turner and Ferrante 1979). However, it should be considered that such measurements include inevitably the nutritional contents of microorganisms which colonize the fecal pellets (Newell 1965; Turner 1979). Moreover, the peritrophic envelope of the fecal pellets not only consists of chitin microfibrils but also of a protein-containing matrix which might increase its nutritional value. The large fecal pellets of salps show the highest sinking rates, between 320 to 2238 m/day. Under these circumstances microbial colonization during sinking seems to be very low.

However, the flocculent fecal masses of *Thalia democratica* and *Dolioletta gegenbauri*, which have a low sinking rate, were colonized by a succession of bacteria and protozoa before they reached deeper waters (Pomeroy and Deibel 1980). Peduzzi and Herndl (1986) investigated the role of bacteria in the decomposition of the fecal pellets of the epiphyte-grazing gastropod *Gibbula umbilicaris*.

A different picture was obtained by Gowing and Silver (1983) who not only investigated the easily accessible surface of fecal pellets by scanning electron microscopy, but also regarded the bacteria in the interior of the pellets by transmission electron microscopy. The results show that microbial decomposition of fecal material is initiated in the sea from inside the fecal pellets where large amounts of bacteria are present. These bacteria are either enteric bacteria or ingested, digestion-resistant bacteria, or both which live under reduced oxygen tension. Therefore, they are quite different from those bacteria which are observed to accumulate on the surface of the pellets in the laboratory. These live in an oxygen-rich environment. Microelectrode studies demonstrated that inside large macro-crustacean fecal pellets oxygen can be completely depleted in the dark or after about 60 h in the laboratory (Alldredge et al. 1987). Outside these pellets the oxygen content increased with the distance from its surface. Gowing and Silver (1983) found that fecal pellets held in the laboratory were covered by dividing bacteria, whereas only few bacteria were seen in division on pellets obtained from the sea. Bacterial biomass of pellets presumably deriving from copepods and larvaceans collected and preserved in the sea amounted to $0.7 \, mm^3 \times 10^{-3}$ within the fecal pellet and $0.02 \, mm^3 \times 10^{-3}$ on its surface. The bacterial biomass of pellets of a swimming crab, *Pleuroncodes planipes*, from the sea was determined as inside $2.6 \, mm^3 \times 10^{-3}$ and on the surface of $0.02 \, mm^3 \times 10^{-3}$. After 80 h in the laboratory the relation was $16 \, mm^3 \times 10^{-3}$ inside and $6 \, mm^3 \times 10^{-3}$ on the surface. This shows that all determinations conducted under laboratory conditions may be wrong as the conditions are unnatural. During incubation in the laboratory, the pellets lie on the bottom of containers onto which bacteria and organic matter are readily adsorbed. The real conditions in the sea cannot be simulated as it is impossible to construct an apparatus which allows free fall of pellets and appropriate temperature conditions.

Fungi did not appear to degrade the peritrophic envelope of fecal pellets of an amphipod, the freshwater detritivore *Gammarus lacustris limnaeus*, although germinating fungal spores were occasionally seen on the surface of these pellets (Lautenschlager et al. 1978).

6.7.10 Coprophagy

Coprophagy by zooplankton and colonization by microorganisms are the most important mechanisms for the recycling of the nutritive contents of fecal pellets. Paffenhöfer and Knowles (1979) could show that it is possible to rear 57–71% of an initial population of a copepod, *Calanus helgolandicus*, from the copepodid stage III to adults only on a diet of fecal pellets. Moreover, they could show that coprophagy takes place in copepods, as intact fecal pellets of nauplii of *Eucalanus pileatus* were ingested by late copepodid stages and adults as well as adults of another copepod species, *Temora stylifera*. These pellets were fed at rates comparable to those of large diatomes. Therefore, these authors suggested that small fecal pellets are recycled in the surface layers, whereas larger pellets due to their higher sinking rate transported organic matter into deeper layers of oceans and lakes. Recycling should be accelerated in the warmer surface waters of subtropical and tropical areas by microbial degradation of the peritrophic envelope of the pellets. Paffenhöfer and Knowles discussed still another interesting aspect. In oceanic areas where often nanoplankton dominates, smaller copepods are able to utilize more efficiently the nanoplankton than the larger species of copepods. The latter could survive in such areas by feeding on the fecal pellets of the smaller species.

Analysis of sinking, resuspension and degradation rates indicate that small fecal pellets must to be heavily grazed in the water column (Emerson and Roff 1987).

Coprophagy is not at all restricted to copepods. Frankenberg and Smith (1967) investigated the rates of coprophagy in 38 species (polychaets, gastropods, bivalves, crustaceans and teleost fish). These were fed fecal pellets from corresponding groups. Fecal ingestion rates were positively correlated with the carbon and nitrogen contents of the pellets. Coprophagy has also been demonstrated in ostracods, euphausiids, mysids, pteropods and salps. Silver and Bruland (1981) demonstrated that salps and pteropods consume small fecal pellets from small crustaceans or larval forms along with phytoplankton. This could result in an unexpected recycling as well as a speeding of the descent of smaller pellets. The smaller fecal pellets were often found intact within the larger pellets of salps and pteropods. Newell (1965) found that the feces of two marine deposit feeders, the prosobranch gastropod *Hydrobia ulvae*, and the bivalve, *Macoma balthica*, had initially a high carbon content. But when these feces had been colonized by microorganisms for 3 days at 19 °C the carbon content was reduced only slightly, whereas the nitrogen contents had increased 40–85 times. Although the feces were not initially accepted by these mollusks, they were readily ingested after such degradation by microorganisms. Therefore, Darnell (1967) assumed that many marine and freshwater detritus feeders ingest fecal pellets primarily on account of the microorganisms which degrade them. This hypothesis is supported by the results of Hargrave (1970) who worked with amphipods, Lopez et al. (1977) who investigated another amphipod, *Orchestia grillus*, Kofoed (1975) who worked with the gastropod, *Hydrobia ventrosa*, and Fenchel (1970) and Levinton (1972) with different species of detritovores. Hansen (1978) used a Phillipson microcomb calorimeter to determine the nutritive value of substrates which were ingested by substrate feeders, *Sipunculus nudus*, *Phascolosoma vulgare* and *Leptosynapta inhaerens*. The largest part of the calorific contents of the substrate was provided by fecal pellets with 0.6 cal/mg dry weight. This accounted for 92% of the total caloric content of the food substrate. The author assumed that peritrophic membranes with attached microorganisms and coprophagy might play an important role.

A peculiar utilization of fecal material has been observed by Youngbluth (1982) in the pelagic mysis larva of the penaeid shrimp *Solenocera atlantidis*. The larvae were observed at night from a submersible at depths ranging from 30 to 300 m near the Bahamas. A strand of fecal material protruded from the anus often five times the body length of a larva. The dangling strand accumulated a considerable amount of living plankton organisms as well as detritus. Periodically the larva flexed its pereion, forced the strand upward and fed on the adhering plankton organisms.

Behavioral, metabolic and biochemical data indicate that the fecal strand with the adhering microbes, algae, other planktonic organisms and organic debris serves as a concentrated food resource for the larvae.

Another exceptional use of feces has been reported by McCloskey (1970) in a small gammaridean amphipod, *Dulichia rhabdoplastis*, which lives together with the sea urchin *Strongylocentrotus franciscanus*. The amphipod constructs a string of fecal pellets, attaches it to the spines of the sea urchin and farms the microorganisms which attach to the string.

6.7.11 Coprorhexy

A new degradation factor for fecal pellets has been introduced. Experimental evidence showed clearly that copepods are able to switch from their well-known herbivorous or filter feeding to carnivorous feeding (Landry 1981). They are also able to break up fecal pellets which was called coprorhexy (from Greek kopros = feces, rhexis = to break or crumble) (Lampitt et al. 1990) in order to discriminate it from coprophagy, the ingestion of fecal pellets (Fig. 78). Copepods were allowed to clear their guts in filtered seawater for at least 2 h and were then transferred for 24 to 30 h to bottles with filtered seawater and different amounts of fecal pellets. Scanning electron microscopic control of such pellets revealed that the copepods had peeled off the peritrophic envelope from many pellets, probably in order to obtain the adhering bacteria. Once the pellet envelope has been removed, ciliates are able to break up the remains considerably.

Alldredge et al. (1987) reported that 40% of the large pellets of euphausiids in surface waters off southern California had peritrophic membranes which were either partially or totally decayed.

6.7.12 Sediment Formation

Fecal pellets are common in marine as well as freshwater bottom sediments. Brundin (1949, cited in Hargrave 1976) observed that sediments in several Swedish lakes were largely composed of chironomid fecal pellets (Diptera, Insecta). Haven and Morales-Alamo (1967) reported that preliminary studies in the York River, Virginia (USA) indicated that in certain localities up to 25% by weight of the sediments may be composed of pellets. Moore (1931a) found that even 30 to 50% of the mud in the Clyde Sea area could be made up of fecal pellets. The greatest and most important changes took place in the uppermost layers, within a few centimeters of the surface. The average of a series of counts showed about 3400 pellets from a maldanid worm (Polychaeta) per cm^3 of mud. The fecal pellets formed by these worms remain more or less unchanged in the bottom deposits for periods of 50 to 100 years and might be eventually mineralized.

As normally most of the sedimenting material has a size of less than 40 μm, the faster settling rates of the more compact fecal pellets might accelerate the sedimentation of the finer, more slowly settling materials (Haven and Morales-Alamo 1967). Taghon et al. (1984) pointed out that transport and breakdown of fecal

Fig. 78a-c. Scanning electron micrographs of fecal pellets of marine copepods (after Lampitt et al. 1990). *a* Fecal pellet with intact peritrophic envelope; 240x. *b* After some days the peritrophic envelope begins to decay; 425x. *c* Copepods are able to remove the peritrophic envelope of fecal pellets with their mouthparts. On the *right side*, an egg (*E*) of *Centropages hamatus* is attached to the fecal pellet; 450x

pellets have sedimentological conseqences. These authors could show that pellets of a polychaete, *Amphicteis scaphobranchiata*, living subtidally in Puget Sound (Washington, USA) and down to 2000 m off the coast of southern California, were transported for variable distances depending on their age. Thus, older pellets (6 h after production) traveled a median distance of 3 m, whereas freshly voided pellets traveled 9.5 m before disintegrating. Generally, pellets of different morphologies, produced by different species, have increased or decreased critical shear stresses for movement as compared with unpelletized sediment.

6.7.13 Element Depletion in Surface Layers and Recycling Problems

During blooms of phytoplankton, the surface layers are depleted of certain essential elements, as fecal pellets transfer these substances into deeper water or to the floor of oceans or lakes. In this way large quantities of phosphorus or silica are removed rapidly from the upper layers. Recycling takes different times and the mechanisms of recycling are unknown. A special problem, or paradox as Honjo (1976) called it, is the fact that sediments composed of calcareous skeletons of planktonic coccolithophorids or siliceous skeletons of planktonic diatoms are deposited below water which is undersaturated with either calcite or silica. Calcite or silica should dissolve even in cold water and under high pressure. Fecal pellet membranes seem to be the protective mechanism which allows recycling only gradually. The rate at which silica is deposited in this way exceeds the flux of dissolved silica and seems to be responsible for the undersaturation of seawater with silica (Schrader 1971).

6.7.14 Terrestrial Ecosystems

Coprophagy not only occurs in aquatic but also in terrestrial ecosystems. Wieser (1965, 1966) could show that under cultural conditions the isopod *Porcellio scaber* must ingest fecal pellets to obtain sufficient copper needed in its metabolism.

6.8 Peritrophic Membranes Used as Cocoon Material for Pupation

Many insect larvae spin a cocoon before pupation. The thread consists of protein which is secreted by labial glands, Malpighian tubules or glands on thorax or abdomen. Schulze (1927) reported that the larvae of Javanese Tenebrionidae, *Platydema tricuspis*, which feed on fungi of trees, produce a chitin-containing thread. But the chitin derives from undigested cell walls of fungi which are glued together by secretions formed by the ileum.

Several species of Coleoptera from different groups use peritrophic membranes as cocoon material:

Order Coleoptera
    Bostrychiformia
        Family Ptinidae
            Ptinus tectus
            Niptus hololeucus
            Gibbium psylloides
            Mezium affine
            M. americanum
        Family Cerylonidae
            Murmidius ovalis
    Cucujiformia
        Family Curculionidae
            Rhynchaenus fagi
            Prionomerus calceatus
            Cionus scrophulariae
            Cleopus pulchellus

Streng (1969, 1973) was the first to detect that the chitin-containing cocoon material of a small weevil larva, *Rhynchaenus fagi* (Curculionidae), is not secreted by the Malpighian tubules as had been generally assumed, but in the midgut. These larvae feed as miners on the leaves of beeches. The third and last larvae feed almost continuously for 2 days and therefore are called feeding larvae. Peritrophic membranes are secreted by the midgut epithelium of these larvae which contain microfibrils arranged in a hexagonal or orthogonal texture. After about 2 days the larva stops feeding and wanders about in the mine for about 36 h. During this time, the contents of the midgut are voided. Then the larva comes to rest and starts with the formation of a cocoon. In the so-called spinning larvae the posterior part of the midgut is enlarged (Fig. 79). Peritrophic membranes are delaminated synchronously in its anterior part. Additionally, chitin fibrils are secreted in the anterior midgut. These are passed into the posterior midgut and surrounded by the peritrophic membranes. Both materials are transformed into a solid and sclerotized thread in the thread gut (Fig. 79) which is present in spinning larvae only. The thread gut forms the thread and transports it by vigorous peristalsis into the reservoir where up to 30 cm of thread may be accumulated. At 21 °C about 10 mm of thread are produced per minute. A larva needs about 11 m of thread for the formation of its cocoon.

Rudall and Kenchington (1971, 1973) discovered independently in the Australian spider beetle, *Ptinus tectus*, that the cocoon material is not produced by the Malpighian tubules as had been assumed by Marcus (1930), but that it consisted of peritrophic membranes originating from the midgut. Similarly, cocoon material is formed in the posterior midgut of larvae of other species of weevils.

The larvae of *Prionomerus calceatus* are leaf miners. Their feces are surrounded by a continuous, tube-like peritrophic membrane which is trailed behind

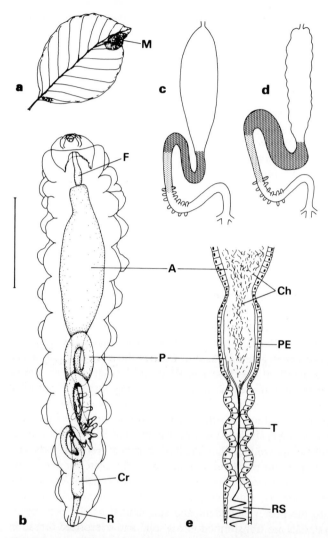

Fig. 79a-e. Formation of a chitin-containing cocoon thread from peritrophic membranes in the larva of *Rhynchaenus fagi* (Coleoptera, Curculionidae). *a* Larva of *Rynchaenus fagi* mining (*M*) near the tip of a beech leaf. *b* Survey of the alimentary canal of the last larval instar seen from the ventral side. *A* Anterior part of the midgut; *Cr* cryptonephridial part of the hindgut; *F* foregut; *P* posterior part of the midgut; *R* rectum. *c,d* Schematic representation of the midgut of a larva during the feeding stage (*c*) and spinning stage (*d*). The region of membrane formation is indicated by *cross-hatching*, the region of thread formation is *stippled*. *e* Schematic representation of the formation of cocoon thread. *Ch* Chitin-containing fibers; *PE* peritrophic envelope; *RS* reservoir with up to 30 cm of usable thread (*T*). About 11 m of thread are needed for the formation of a cocoon. *b-e* After Streng (1973)

the larvae. In the beginning of pupation the first part of the cocoon thread consists of such feces-containing peritrophic membranes. But empty peritrophic membranes are used for the bulk of the cocoon thread. The cocoon thread of this species is the least modified of the peritrophic membranes investigated. All peritrophic membrane material shows a mainly orthogonal texture of chitin microfibrils. X-ray diffraction proved that the γ-type of chitin is present. If the dry cocoon thread is put into water, the original tube form is resumed (Kenchington 1976).

Larvae of *Cionus scrophulariae* and *Cleopus pulchellus* are covered on the dorsal and lateral surfaces with a mucilaginous substance which exudes from the anus and is spread forward by peristaltic movements (Perris 1850). This author observed already that the substance was produced in larger quantities prior to pupation. Prell (1925) noted that the feeding larva is covered by slime and that the cocoon material is secreted by midgut cells. Tristram (1978) gave a detailed description of the formation of peritrophic membranes in these species. The midgut of feeding larvae consists of a bulbous anterior part and a slightly longer and thinner posterior region. A slight constriction marks the junction between these parts. The microvilli of the epithelial cells are rather short, being 2.5 μm long. Secretion of a rather tenuous peritrophic membrane might begin at the end of the anterior midgut, but Tristram emphasized that a substantial structure can be seen only at the beginning of the posterior midgut. Therefore, in these species the secretion of peritrophic membrane seems to be restricted to the posterior midgut. In this case no peritrophic envelope made up of several peritrophic membranes is formed, but a rather fragile gelatinous structure which consists of a loose array of microfibrils and particles. The microfibrils are arranged randomly. Rudall and Kenchington (1973) and Tristram (1978) observed that the production of normal peritrophic membranes ceases when the larvae are fully grown; then the midgut is modified. Its diameter increases by about 25%. Beginning at the constriction, which remains visible, the appearance of the posterior midgut epithelium changes abruptly. The cells are columnar and have microvilli which are about 1 μm long. They are separated by crypts about 60 μm deep in which the cells have microvilli which are 9 mm long (Fig. 80). Already 30–60 min after the larvae cease to feed the cells of the crypts start to secrete large numbers of minute, chitin-containing ribbons. These ribbons on average measure about 75 × 15 μm (Fig. 80). The histological details of ribbon formation do not differ in the larvae of the three species investigated, *Cionus scrophulariae*, *C. hortulanus*, *C. alauda* (Rudall and Kenchington 1973; Kenchington 1976). The chitin-containing microfibrils are arranged randomly. The authors assumed that the ribbons are formed as aggregations of microfibrils near the tips of the microvilli and that only a single ribbon is formed at a time inside a crypt. The ribbons are stored in the lumen of the posterior midgut. During cocoon formation, the ribbons are voided at intervals as a viscous mass and spread over the entire surface of the larva as described by Perris (1850) and Prell (1925). The ribbons seem to stick to the surface of the larvae by the slime produced already at the feeding stage. Ribbons can be separated from each other if freshly extruded material is immersed in water and shaken vigorously or ultrasonicated. Hardened cocoons must be deproteinized with either hot alkali or SDS before individual ribbons can be isolated. Their size varies considerably, but on an average they are 70 μm long and about 10 μm wide.

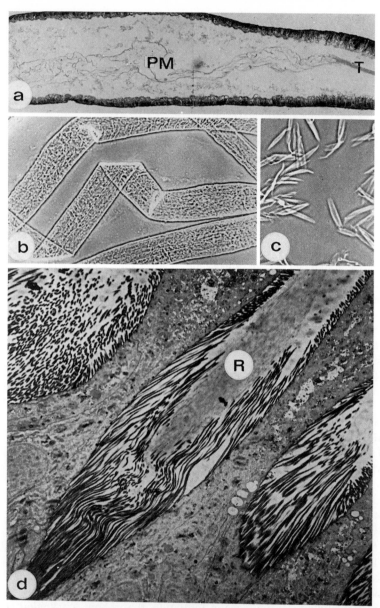

Fig. 80a-d. Use of peritrophic membranes as cocoon silk for pupation in some Coleoptera (after Rudall and Kenchington 1973). *a* A longitudinal section of the midgut of the larva of *Gibbium psylloides* shows how the peritrophic membranes (*PM*) are transformed into a silk thread (*T*). Stained with Heidenhain's Azan; 100x. *b* Isolated, flat, ribbon-like silk threads from a larval cocoon of the same species; 100x. *c* Isolated ribbons from a cocoon of *Cionus scrophulariae*. Phase contrast; 100x. *d* A section of the midgut of a larva of *Cleopus pulchellus* shows the deep intercellular crypts. In one of these a ribbon (*R*) has been secreted; 4100x

Marcus (1930) investigated the alimentary canal of the larvae of *Niptus hololeucus* and *Ptinus fur* histologically and assumed that the cocoon silk was secreted by the Malpighian tubules. However, electron microscopic investigations by Rudall and Kenchington (1973) revealed that in *Ptinus tectus* the cocoon thread is secreted by the midgut epithelium and consists of slightly modified peritrophic membranes. In feeding larvae of *P. tectus* and *Gibbium psylloides* (Tristram 1977) the midgut is a more or less straight tube which is about 9 mm long and can be divided into two parts of similar length, the dilated anterior or bulbous midgut and the posterior or thin midgut. In both species the peritrophic envelope is not secreted by the whole midgut, but in a restricted part of it, in the middle region. The secretion of peritrophic membranes starts in the latter part of the anterior midgut and more membranes are added by the epithelium of the first 1–2 mm of the posterior midgut epithelium. This region is characterized by a columnar epithelium in which the cells show a height:width ratio of about 5:1. The remaining cells of the midgut epithelium are more or less cubic. The peritrophic envelope consists usually of four to five peritrophic membranes which form a concentric tube. The chitin microfibrils are arranged in a random, hexagonal or orthogonal texture, the random form being more common in those membranes which are used for the construction of the cocoon. A regular network of microfibrils appears adjacent to the tips of the microvilli but cannot be seen between the microvilli. The microfiber bundles have a width of about 50 nm. In regular textures the bundles are about 0.2 μm apart. Again, the fully grown larva ceases to feed and empties its gut. The anterior midgut collapses and is reconstructed from nidi of replacement cells which form the new adult epithelium. Meanwhile, the formation of peritrophic membranes continues in the posterior midgut which seems to persist. The tube-like peritrophic envelope is transformed in the gut into a flat ribbon (Fig. 80), but there is not a distinct press as in *Rhynchaenus fagi*. However, Tristram observed longitudinal and lateral ridges which seem to be caused by compression in the hindgut and by the bristle-edged anal teeth, respectively. The ribbons are stored in the posterior midgut but they seem to be folded to a lesser extent than those of *Rhynchaenus fagi*. The ribbons, which vary in width from 20–100 μm, are extruded from the anus. The microfibrils tend to aggregate in *P. textus*. X-ray diffraction showed that γ-chitin is present in the membranes of both species. The cocoon threads of *Mezium affine* and *M. americanum* are very similar in appearance to those of *G. psylloides* (Rudall, unpubl. observ.).

In other species of the Ptinidae, *Niptus hololeucus*, *Pseudorostus hilleri*, *Ptinus sexpunctatus* and *Stothomezium squamosum* the cocoon consists of pieces of wood and detritus. If cocoons of *N. hololeucus* were immersed in water a sparse amount of viscous material was observed which might act as an adhesive to stick the particles together. No thread-like material derived from more or less modified peritrophic membranes seems to be present. Therefore, it was assumed that this material may be secreted by the Malpighian tubules (Kenchington 1976).

In feeding larvae of *Murmidius ovalis* peritrophic membranes could not be detected. The midgut has grossly the same morphology as *Cionus scrophulariae*. The formation of ribbons for the construction of a cocoon has been investigated only incompletely (Kenchington 1976). Ribbons seem to be formed from fibrous material. In electron micrographs the microfibrils appear already between the bases of the microvilli. They are considerably longer and usually thinner than those of *Cionus scrophulariae*, about 200 × 8 μm. Each end is tapered to a fine point. X-ray diffraction has shown that chitin occurs in the β-form.

There are probably more examples among beetles, even in other groups than those mentioned above, in which peritrophic membranes are used for the formation of cocoons.

References

Abbot RT (1954) Review of the Atlantic periwinkles, *Nodilittorina, Echinius* and *Tectarius.* Proc US Natl Mus 103:449–464

Abedi ZH, Brown AWA (1961) Petritrophic membrane as a vehicle for DDT and DDE excretion in *Aedes aegypti* larvae. Ann Entomol Soc Am 54:539–542

Adam H (1960) Über das Vorkommen einer peritrophischen Membran im Darm von *Myxine glutinosa* L. (Cyclostomata). Naturwissenschaften 47:333–334

Adam H (1963) Structure and histochemistry of the alimentary canal. In: Brodal A, Fänge R (eds) The biology of *Myxine.* Univ Press, Oslo, pp 256–288

Adam H (1965) Einige Beiträge zur Kenntnis des Darmes von *Myxine glutinosa* L. (Cyclostomata). Verh Dtsch Zool Ges 1964 Kiel; Zool Anz Suppl 28:311–319

Adang MJ, Spence KD (1981) Surface morphology of peritrophic membrane formation in the cabbage looper *Trichoplusia ni.* Cell Tissue Res 218:141–147

Adang MJ, Spence KD (1982) Biochemical comparisons of the peritrophic membrane of the lepidopterans *Orgyia pseudotsugata* and Manduca sexta. Comp Biochem Physiol 73B:645–649

Adang MJ, Spence KD (1983) Permeability of the peritrophic membrane of the Douglas fir tussock moth (*Orgyia pseudotsugata*). Comp Biochem Physiol 75A:233–238

Adlerz G (1890) Om digestionssekretionen jeinte några dermed sammanhaengende fenomen hos Insekter och Myriopoder. Bihang Svensk Vet Akad 16 Afd IV 2:51 pp

Agudelo-Silva F, Spielman A (1985) Penetration of mosquito midgut wall by sheathed microfilariae. J Invertebr Pathol 45:117–119

Akimov IA (1980) Functions of gut sections of acaroid mites (Acaroidea). Vestn Zool 1980:17–24

Alldredge Alice L, Cohen Y (1987) Can microscale chemical patches persist in the sea? Microelectrode study of marine snow, fecal pellets. Science NY 235:680–691

Alldredge Alice L, Gottschalk CC, MacIntyre S (1987) Evidence for sustained residence of macrocrustacean fecal pellets in surface waters off southern California. Deep-Sea Res 34:1641–1652

Allen K, Neuberger A, Sharon N (1973) The purification, composition and specificity of wheat germ agglutinin. Biochem J 131:155–162

Ameen M, Rahman MF (1973) Larval and adult digestive tracts of *Tribolium castaneum* (Herbst) (Coleoptera:Tenebrionidae). Int J Insect Morphol Embryol 2:137–152

Andrews CC, Karl DM, Small LF, Fowler SW (1984) Metabolic activity and bioluminescence of oceanic faecal pellets and sediment trap particles. Natufe (London) 307:539–541

Andries J-C, Torpier G (1982) An extracellular brush border coat of lipid membranes in the midgut of *Nepa cinerea* (Insecta, Heteroptera): ultrastructure and genesis. Biol Cell 46:195–202

Angel MV (1984) Detrital organic fluxes through pelagic ecosystems. In: Faham MI (ed) Flows of matter and energy in marine ecosystems: theory and practice. Plenum, New York, pp 475–516

Anglas J (1901) Observations sur les métamorphoses internes de la guêpe et de l'abeille. Bull Sci Fr Belg 34:363–473

Aronson M, Medalia O, Schori L, Mirelman D, Sharon N, Ofek I (1979) Prevention of colonization of the urinary tract of mice with *Escherichia coli* by blocking of bacterial adherence with methyl-α-D-manno-pyranoside. J Infect Dis 139:329–332

Asakura A (1982) A possible role of the gastric caecum in osmoregulation of the seawater mosquito *Aedes togoi.* Annones Zool Jpn 55:1–8

Aschner M (1931) Die Bacterienflora der Pupiparen (Diptera). Eine Symbiosestudie an blutsaugenden Insekten. Z Morphol Ökol Tiere 20:368–442

Aubertot M (1932a) Origine proventriculaire et évacuation continue de la membrane péritrophique chez les larves d'*Eristalis tenax*. C R Soc Biol 111:743–745

Aubertot M (1932b) Les sacs péritrophiques des larves d'*Aeschna*; leur évacuation périodique. C R Soc Biol 111:746–748

Aubertot M (1934) Recherches sur les membranes péritrophiques des insectes et en particulier des Diptères. Thesis, Univ Strasbourg, Nancy (E) 44

Baerwald RJ, Delcarpio JB (1983) Double membrane-bounded intestinal microvilli in *Oncopeltus fasciatus*. Cell Tissue Res 232:593–600

Bain O, Philippon B (1970) Mécanisme de la traversée de la paroi stomacle par les microfilaires chez *Anopheles stephensi* et *Simulium damnosum*. Ann Parasit Human Comp 45:295–320

Bain O, Philippon B, Séchan Y, Cassone J (1976) Corrélation entre le nombre de microfilaires ingérées et l'épaisseur de la membrane péritrophique du vecteur dans l'Onchocercose de savane africaine. C R Acad Sci Paris D 283:391–392

Baines DM (1978) Observations on the peritrophic membrane of *Locusta migratoria migratoroides* (R. and F.) nymphs. Acrida 7:11–21

Baird RC, Thompson NP, Hopkins TL, Weiss WR (1975) Chlorinated hydrocarbons in mesopelagic fishes of the eastern Gulf of Mexico. Bull Mar Sci 25:473–481

Balbiani EG (1890) Études anatomiques et histologiques sur le tube digestif des *Cryptops*. Arch Zool Exp Gen 8:1–82

Bamrick JF (1964) Resistance to American foulbrood in honey-bees. 5. Comparative pathogenesis in resistant and susceptible larvae. J Insect Pathol 6:284–304

Bandel K (1974) Faecal pellets of Amphineura and Prosobranchia (Mollusca) from the Caribbean Coast of Columbia, South America. Senckenberg Mar 6:1–31

Barrett FM (1982) Absorption of fluid from the anterior midgut of *Rhodnius prolixus*. J Insect Physiol 28:335–341

Batham EJ (1945) *Pollicipes spinosus* Quoy and Gaimard. I. Notes on biology and anatomy of adult barnacle. Trans Proc R Soc NZ 74:359–374

Bathelt RW, Schelske CL (1983) Degradation of the peritrophic membrane of fresh-water zooplankton fecal pellets. Trans Am Microsc Soc 102:288–299

Bathmann U, Liebezeit G (1986) Chlorophyll in copepod faecal pellets: changes in pellet numbers and pigment content during a declining Baltic spring bloom. PSZN Mar Ecol 7:59–73

Bathmann UV, Noji TT, Voss M, Peinert R (1987) Copepod fecal pellets: abundance, sedimentation and content at a permanent station in the Norwegian Sea in May/June 1986. Mar Ecol Prog Ser 38:45–51

Bauer PG (1981) Ultrastrukturelle und physiologische Aspekte des Mitteldarms von *Rhodnius prolixus* Stal (Insecta, Heteroptera). Thesis, Univ Basel, 125 pp

Beachy EH, Offek I (1976) Epithelial cell binding of group A streptococci by lipoteichoic acid on fimbriae denuded of M protein. J Exp Med 143:759–771

Beasley TM, Heyraud M, Higgo JJW, Cherry RD, Fowler SW (1978) [210]Po and ^{210}Pb in zooplankton fecal pellets. Mar Biol 44:325–328

Becker A, Peters W (1985) The ultrastructure of the midgut and the formation of peritrophic membranes in a harvestman, *Phalangium opilio* (Chelicerata, Phalangida). Zoomorphology 105:326–332

Becker B (1977) Licht- und elektronenmikroskopische Untersuchungen zur Bildung peritrophischer Membranen bei der Imago von *Calliphora erythrocephala* Meig. (Insecta, Diptera). Zoomorphologie 87:247–262

Becker B (1978a) Determination of the formation rate of peritrophic membranes in some Diptera. J Insect Physiol 24:529–533

Becker B (1978b) Effects of 20-hydroxy-ecdysone, juvenile hormone, Dimilin, and Captan on in vitro synthesis of peritrophic membranes in *Calliphora erythrocephala*. J Insect Physiol 24:699–705

Becker B (1980) Effects of Polyoxin D on in vitro synthesis of peritrophic membranes in *Calliphora erythrocephala*. Insect Biochem 10:101–106

Becker B, Peters W, Zimmermann U (1975) Investigations on the transport function and structure of peritrophic membranes. VI. In vitro synthesis of peritrophic membranes of the blowfly, *Calliphora erythrocephala*. J Insect Physiol 21:1463–1470

Becker B, Peters W, Zimmermann U (1976) Investigations on the transport function and structure of peritrophic membranes. VII The fine structure of peritrophic membranes of the blowfly, *Calliphora erythrocephala*, grown in vitro under different conditions. J Insect Physiol 22:337–345

Benayoun G, Fowler SW, Oregioni B (1974) Flux of cadmium through euphausiids. Mar Biol 27:205–212

Benjaminson MA (1969) Conjugates of chitinase with fluorescein isothiocyanate or lissamine rhodamine as specific stains for chitin in situ. Stain Technol 44:27–31

Berlese A (1909) Gli Insetti. I. Embriologia e morfologia. Societa editrice libraria, Milano

Bernays EA (1981) A specialized region of the gastric caeca in the locust, *Schistocerca gregaria*. Physiol Entomol 6:1–6

Bernays EA, Chamberlain DJ (1980) Tolerance of ingested tannin in *Schistocerca gregaria*. J Insect Physiol 26:415–420

Bernays EA, Chamberlain DJ, McCarthy P (1980) The differential effects of ingested tannic acid on different species of Acridoidea. Entomol Exp Appl 28:158–166

Berner R, Rudin W, Hecker H (1983) Peritrophic membranes and protease activity in the midgut of the malaria mosquito, *Anopheles stephensi* (Liston) (Insecta: Diptera) under normal and experimental conditions. J Ultrastruct Res 83:195–204

Berridge MJ (1970) A structural analysis of intestinal absorption. Symp R Entomol Soc Lond 5:135–150

Bertram DS, Bird RG (1961) Studies on mosquito-borne viruses in their vectors. 1. The normal fine structure of the midgut epithelium of the adult female *Aedes aegypti* (L.) and the functional significance of its modifications following a blood meal. Trans R Soc Trop Med Hyg 55:404–423

Bess HA (1935) The alimentary canal of *Calosoma sycophanta* Linnaeus. Ohio J Sci 35:97–105

Bienfang PK (1980) Herbivore diet affects fecal pellet settling. Can J Fish Aquat Sci 37:1352–1357

Bignell DE (1981) Comments on functions of the peritrophic membrane in haematophagous insects. Parasitology 82:95–97

Bignell DE, Oskarsson H, Anderson JM (1980) Colonization of the epithelial face of the peritrophic membrane and the ectoperitrophic space by Actinomycetes in a soil-feeding termite. J Invertebrate Pathol 36:426–428

Billingsley PF, Downe AER (1983) Ultrastructural changes in posterior midgut cells associated with blood-feeding in adult female *Rhodnius prolixus* Stal (Hemiptera: Reduviidae). Can J Zool 61:1175–1187

Bishop JKB, Edmond JM, Ketten DR, Bacon MP, Silker WB (1977) The chemistry, biology and vertical flux of particulate matter from the upper 400 m of the equatorial Atlantic. Deep Sea Res 24:511–548

Bishop JKB, Ketten DR, Edmond JM (1978) The chemistry, biology and vertical flux of particulate matter from the upper 400 m of the Cape Basin in the southeast Atlantic. Deep Sea Res 25:1121–1161

Bodungen B von (1986) Phytoplankton growth and krill grazing during spring in the Bransfield Strait, Antarctica implications from sediment trap collections. Polar Biol 6:153–160

Boothe PN, Knauer GA (1972) The possible importance of fecal material in the biological amplification of trace and heavy metals. Limnol Oceanogr 17:270–274

Bordas L (1894) Anatomie du tube digestif des Hyménoptères. C R Acad Sci Paris 118:1423

Bordas L (1911) L'appareil digestif et les tube de Malpighi des larves des Lépidoptères. Ann Sci Natl Zool 14:191–273

Borght O van der (1966) Peritrophic membranes in Arachnida (Arthropoda). Nature (Lond) 210:751–752

Borovsky D (1986) Proteolytic enzymes and blood digestion in the mosquito *Culex nigripalpus*. Arch Insect Biochem Physiol 3:147–160

Bowen RC (1968) Histochemical studies on two millipede species. Ohio J Sci 68:85–91

Bowen VT, Robinson AC, Sutton D (1951) The uptake and distribution of [140]barium and [140]lanthanum in larvae of *Drosophila repleta*. J Exp Zool 118:509–530

Bracke JW, Cruden DL, Markovetz AJ (1979) Intestinal microbial flora of the American cockroach, *Periplaneta americana* L. Appl Environ Microbiol 38:945–955

Bradford M (1976) A rapid and sensitive method for the quantitation of microgram quantities of protein utilizing the principle of protein-dye-binding. Anal Biochem 72:248–254

Bradley TJ (1987) Physiology of osmoregulation in mosquitoes. Annu Rev Entomol 32:439–462

Brandt C, Adang M, Spence KD (1978) The peritrophic membrane: ultrastructural analysis and function as a mechanical barrier to microbial infection in *Orgyia pseudotsugata*. J Invertebrate Pathol 32:12–24

Briegel H, Lea AO (1975) Relationship between protein and proteolytic activity in the midgut of mosquitoes. J Insect Physiol 21:1597–1604

Brooks MA (1963) Insect pathology. An advanced treatise. Academic Press, New York

Browning J (1979) Octopus microvasculature: permeability to ferritin and carbon. Tissue Cell 11:371–383

Bruland KW, Silver WM (1981) Sinking rates of fecal pellets from gelatinous zooplankton (salps, pteropods, doliolids). Mar Biol 63:295–300

Burgos MH, Gutiérrez LS (1976) The intestine of *Triatoma infestans*. I. Cytology of the midgut. J Ultrastruct Res 57:1–9

Burnette WN (1981) "Western blotting": electrophoretic transfer of proteins from sodium dodecyl sulfate-polyacrylamide gels to unmodified nitrocellulose and radiographic detection with antibody and radioiodinated Protein A. Anal Biochem 112:195–203

Bursell A, Berridge MJ (1962) The invasion of polymorphic trypanosomes of the ecto-peritrophic space of tsetse flies in relation to some properties of the rectal fluid. Ann Soc Exp Biol, Annu Rep

Campbell FL (1929) The detection and estimation of insect chitin and the unrelatedness of chitinization to hardness and pigmentation of the cuticula of the American cockroach *Periplaneta americana*. Ann Entomol Soc Am 22:401–426

Castelnuovo G (1934) Ricerche istologiche e fisiologiche sul tubo digerente di *Carausius (Dixippus) morosus*. Arch Zool Ital 20:443–466

Caulfield J, Farquhar MG (1974) The permeability of glomerular capillaries to graded dextrans. J Cell Biol 63:883–903

Chabaud A, Bain O, Landau I, Petit G (1986) La transmission des parasites par vecteur hématophages: richesse des phénomènes adaptifs. Vie Sci C R Ser Gen 3:469–484

Chaika SY (1979) The properties of the ultrastructural organization of the midgut of bugs (Hemiptera). Vestn Mosk Univ Biol 16:54–61 (in Russian)

Chaika SY (1980a) Ultrastructure of the midgut of *Ceratophyllus sciurorum* Schrank (Siphonaptera). Biol Nauk (Moscow) 1980:61–64 (in Russian)

Chaika SY (1980b) Ultrastructural organization of the midgut of the caddisfly *Limnephilus stigma* Curtis (Trichoptera, Limnephilidae). Entomol Obozr 59:67–72 (in Russian)

Chatton E (1920) Les membranes péritrophiques des Drosophiles (Diptères) et des Daphnies (Cladocères); leur genèse et leur rôle à l'égard des parasites intestinaux. Bull Soc Zool Ser 45:265–280

Cheeseman MT, Pritchard G (1984) Peritrophic membranes in adults of two species of carabid beetles (Coleoptera: Carabidae). Int J Insect Morphol Embryol 13:171–173

Cherry RD, Fowler SW, Beasley TM, Heyraud M (1975) Polonium-210: its vertical oceanic transport by zooplankton metabolic activity. Mar Chem 3:105–110

Christensen BM (1978) *Dirofilaria immitis*: effect on the longevity of *Aedes trivittatus*. Exp Parasitol 44:116–123

Christensen BM, Sutherland DR (1984) *Brugia pahangi*: exsheathment and midgut penetration in *Aedes aegypti*. Trans Am Microsc Soc 103:423–433

Christophers SR (1960) *Aedes aegypti*: the yellow fever mosquito. Its life history, bionomics and structure. Univ Press, Cambridge

Cioffi M (1984) Comparative ultrastructure of arthropod transporting epithelia. Am Zool 24:139–156

Clarke L, Temple GHR, Vincent JFV (1977) The effects of a chitin inhibitor – Dimilin – on the production of peritrophic membrane in the locust, *Locusta migratoria*. J Insect Physiol 23:241–246

Collins C, Miller T (1977) Studies on the action of biogenic amines on cockroach heart. J Exp Biol 67:1–15

Cosper TC, Reeve MR (1975) Digestive efficiency of the chaetognath *Sagitta hispida*. J Exp Mar Biol Ecol 17:33–38

Crawford CS (1972) Water relations in a desert millipede *Orthoporus ornatus* (Girard) (Spirostreptidae). Comp Biochem Physiol 42A:521–535

Croft SL, East JS, Molyneux DH (1982) Anti-trypanosomal factor in the haemolymph of *Glossina*. Acta Trop 39:293–302

Cruz Landim C (1985) Origin of the peritrophic membrane of the adult *Apis mellifera* L. (Hymenoptera: Apidae). Rev Bras Biol 45:207–219

Cuénot L (1895) Études physiologiques sur les Orthoptères. Arch Biol 14:293–341

Dadd RH (1975) Alkalinity within the midgut of mosquito larvae with alkaline-active digestive enzymes. J Insect Physiol 21:1847–1853

Dales RP (1955) Feeding and digestion in terebellid polychaetes. J Mar Biol Assoc UK 34:55

Dales RP (1967) Annelids, 2nd edn. Hutchinson, London

Dales RP, Pell JS (1970) The nature of the peritrophic membrane in the gut of the terebellid polychaete *Neoamphitrite figulus*. Comp Biochem Physiol 34:819–826

Dall W (1967) The functional anatomy of the digestive tract of a shrimp *Metapenaeus bennettae* Racek and Dall (Crustacea: Decapoda: Penaeidae). Aust J Zool 15:699–714

Dallai R (1974) Ultrastructure of the intestine of Protura. Boll Zool 41:475

Daoust RA, Gunner HB (1979) Synergistic microbial attack on *Lymantria dispar* L. (Lepidoptera: Lymantriidae). J NY Entomol Soc 86:284

Darnell RM (1967) Organic detritus in relation to the estuarine ecosystem. In: Lauf GH (ed) Estuaries. Am Assoc Adv Sci Publ 83, Washington DC, pp 376–382

Darwin C (1851–1854) A monograph of Cirripedia. Ray Society, London

Davidson EW (1979) Ultrastructure of midgut events in the pathogenesis of *Bacillus sphaericus* strain SSII-1 infections of *Culex pipiens quinquefasciatus* larvae. Can J Microbiol 25:178–184

DeBelder AN, Granath K (1973) Preparation and properties of fluorescein labelled dextrans. Carbohydrate Res 30:375–378

Dehn M von (1933) Untersuchungen über die Bildung der peritrophischen Membran bei den Insekten. Z Zellforsch Mikrosk Anat 19:79–105

Dehn M von (1936) Zur Frage der Natur der peritrophischen Membran bei den Insekten. Z Zellforsch Mikrosk Anat 25:787–791

Delachambre J (1969) La réaction de la chitine à l'acide-périodique-schiff. Histochemie 20:58–67

DePriester W (1971) Ultrastructure of the midgut epithelial cells in the fly *Calliphora erythrocephala*. J Ultrastruct Res 36:783–805

Desser SS, Wright KA (1968) A preliminary study of the fine structure of the ookinete, oocyst, and sporozoite formation of *Leucocytozoon simondi* Mathis and Leger. Can J Zool 46:303–307

Detra RL, Romoser WS (1979) Permeability of *Aedes aegypti* larval peritrophic membrane to proteolytic enzyme. Mosquito News 39:582–585

Dinsdale D (1975) Excretion in an oribatid mite Phthiracarus sp. (Arachnida: Acari) J Zool Lond 177:225–231

Dörner R, Peters W (1988) Localization of sugar components of glycoproteins in peritrophic membranes of larvae of Diptera (Culicidae, Simuliidae). Entomol Gen 14:11–24

Dow JAT (1981) Countercurrent flows, water movements and nutrient absorption in the locust midgut. J Insect Physiol 27:579–585

Eguchi M, Iwamoto A (1976) Alkaline proteases in the midgut tissue and digestive fluid of the silkworm *Bombyx mori*. Insect Biochem 6:491–496

Eguchi M, Iwamoto A, Yamauchi K (1982) Interrelation of proteases from the midgut lumen, epithelia and peritrophic membrane of the silkworm, *Bombyx mori* L. Comp Biochem Physiol 72 A:359–363

Elder DL, Fowler SW (1977) Polychlorinated biphenyls: penetration into the deep ocean by zooplankton fecal pellet transport. Science 197:459–461

Ellis DS, Evans DA (1977) Passage of *Trypanosoma brucei rhodesiense* through the peritrophic membrane of *Glossina morsitans morsitans*. Nature (Lond) 267:834–835

Ellis DS, Maudlin I (1985) The behaviour of trypanosomes within the midguts of wild-caught tsetse from Zimbabwe. Trans R Soc Trop Med Hyg 79:867–868

El-Mallakh RS (1984) Cold stress and permeability of muscid gut to bacteria. Entomol Exp Appl 36:296–298

Emerson CW, Roff JC (1987) Implications of fecal pellet size and zooplankton behavior to estimates of pelagic-benthic carbon flux. Mar Ecol Prog Ser 35:251–257

Engel E-O (1924) Das Rectum der Dipteren in morphologischer und histologischer Hinsicht. Z Wiss Zool 122:503–533

Espinoza-Fuentes FP, Ferreira C, Terra WR (1984) Spatial organization of digestion in the larval and imaginal stages of the sciarid fly *Trichosia pubescens*. Insect Biochem 14:631–638

Esslinger JH (1962) Behaviour of microfilariae of *Brugia pahangi* in *Anopheles quadrimaculatus*. Am J Trop Med 11:749–758

Evans DA, Ellis DS (1983) Recent observations on the behaviour of certain trypanosomes within their insect hosts. Adv Parasitol 22:1–42

Evans DA, Ellis DS, Stamford S (1979) Ultrastructural studies on certain aspects of the development of *Trypanosoma congolense* in *Glossina morsitans morsitans*. J Protozool 26:557–563

Evenius J (1925) Die Entwicklung des Zwischendarmes der Honigbiene (*Apis mellifica* L.). Zool Anz 63:49–64

Fallis AM (1964) Feeding and related behaviour of female Simuliidae (Diptera). Exp Parasitol 15:439–470

Fallis AM, Bennett GF (1961) Sporogony of *Leucocytozoon* and *Haemoproteus* in simuliids and ceratopogonids and a revised classification of the Haemosporidia. Can J Zool 39:215–228

Farkas B (1922) Beiträge zur Kenntnis der Anatomie und Histologie des Darmkanals der Copepoden. Acta Univ Szeged Sci Mat 1:47–76

Farmer J, Maddrell SHP, Spring JH (1981) Absorption of fluid by the midgut of *Rhodnius*. J Exp Biol 94:301–316

Fenchel T (1970) Studies on the decomposition of organic detritus derived from the turtle grass *Thlassia testudinum*. Limnol Oceanogr 15:14–20

Ferrante JG, Parker JI (1977) Transport of diatom frustules by copepod faecal pellets to the sediments of Lake Michigan. Limnol Oceanogr 22:92–98

Ferrante JG, Parker JI (1978) The influence of planktonic and benthic crustaceans on silicon cycling in Lake Michigan, USA. Verh Int Verein Limnol 20:324–328

Ferreira C, Terra WR (1980) Intracellular distribution of hydrolases in midgut caeca cells from an insect with emphasis on plasma membrane-bound enzymes. Comp Biochem Physiol 66B:467–473

Ferreira C, Terra WR (1982) Function of midgut caeca and ventriculus: microvilli bound enzymes from cells of different midgut regions of starving and feeding *Rhynchosciara americana* larvae. Insect Biochem 12:257–262

Ferreira C, Ribeiro AF, Terra WR (1981) Fine structure of the larval midgut of the fly *Rhynchosciara americana* and its physiological implications. Insect Physiol 27:559–570

Finke R, Peters W (1989) Effects of 5-hydroxytryptamine on the formation of peritrophic membranes in adult blowflies, *Calliphora erythrocephala* (Diptera: Calliphoridae). Entomol Gen 14:243–254

Fishman L, Zlotkin E (1984) A diffusional route of transport of horseradish peroxidase through the midgut of a fleshfly. J Exp Zool 229:189–195

Fishman L, Primor N, Zlotkin E (1984) The selective permeability of fleshfly midgut to an orally toxic cobra venom cardiotoxin. J Exp Biol 108:441–451

Florkin M (1966) A molecular approach to phylogeny. Elsevier, Amsterdam, 176 pp

Foster GR (1953) Peritrophic membranes in the Caridea (Crustacea, Decapoda). J Mar Biol Assoc UK 33:315–318

Forstner GG, Teichman N, Kalnins V, Forstner G (1973) Intestinal goblet cell mucus: isolation and identification by immunofluorescence of a goblet cell glycoprotein. J Cell Sci 12:585–602

Foster R (1963) Infection of Glossina spp. Wiedemann 1830 (Diptera) and domestic stock with Trypanosoma spp. Gruby 1843 (Protozoa) in Liberia. Ann Trop Med Parasitol 57:383–396

Foster R (1964) An unusual protozoal infection of tsetse flies (*Glossina* Wiedemann 1830 spp.) in West Africa. J Protozool 11:100–106

Fowler SW (1977) Trace elements in zooplankton particulate products. Nature (Lond) 269:51–53

Fowler SW, Fisher NS (1983) Viability of marine phytoplankton in zooplankton fecal pellets. Deep Sea Res 30:963–969

Fowler SW, Small LF (1972) Sinking rates of euphausiid fecal pellets. Limnol Oceanogr 17:293–296

Fowler SW, Heyraud M, Small LF, Benayoun G (1973) Flux of ^{141}Ce through a euphausiid crustacean. Mar Biol 21:317–325

Frankenberg D, Smith KL Jr (1967) Coprophagy in marine animals. Limnol Oceanogr 12:443–450

Frankenberg D, Coles SL, Johannes RE (1967) The potential trophic significance of *Callianassa major* fecal pellets. Limnol Oceanogr 12:113–120

Freeman JC (1973) The penetration of the peritrophic membrane of the tsetse fly by trypanosomes. Acta Trop 30:347–354

Frens G (1973) Controlled nucleation for the regulation of the particle size in monodisperse gold solutions. Nature Phys Sci 241:20–22

Frenzel J (1886) Einiges über den Mitteldarm der Insekten sowie Epithelregeneration. Arch Mikrosk Anat 26:229–306

Freyvogel TA (1980) A propos des relations hôte/parasites entre moustiques et plasmodium. Cah ORSTOM, Ser Entomol Med Parasitol 18:171–176

Freyvogel T, Jacquet C (1965) The prerequisites of the formation of a peritrophic membrane in Culicidae females. Acta Trop 22:148–254

Freyvogel T, Stäubli W (1965) The formation of peritrophic membranes in Culicidae. Acta Trop 22:118–147

Füller H (1965) Dichroitische Anfärbung von Chitin mit Thiazinrot, ein histochemischer Chitinnachweis. Zool Anz 174:125–131

Gander E (1968) Zur Histochemie und Histologie des Mitteldarmes von *Aedes aegypti* und *Anopheles stephensi* in Zusammenhang mit der Blutverdauung. Acta Trop 25:133–175

Gansen-Semal P van (1962) Structures et fonctions du tube digestif du lombricien *Eisenia foetida* Sav. Thesis, Univ Louvain, 120 pp

Gass RF, Yeates RA (1979) In vitro damage of cultured ookinetes of *Plasmodium gallinaceum* by digestive proteinases from susceptible *Aedes aegypti*. Acta Trop 36:243–252

Gauld DT (1957) A peritrophic membrane in calanoid copepods. Nature (Lond) 179:325–326

Geigy R, Herbig A (1955) Erreger und Überträger tropischer Krankheiten. Verlag Recht und Gesellschaft, Basel

Gemetchu T (1974) The morphology and fine structure of the midgut and peritrophic membrane of the adult female, *Phlebotomus longipes* Parrot and Martin (Diptera: Psychodidae). Ann Trop Med Parasitol 68:111–124

Geoghegan WD, Ackerman GA (1977) Adsorption of horseradish peroxidase, ovomucoid and anti-immunoglobulin to colloid gold for the indirect detection of Concanavalin A, wheat germ agglutinin and goat antihuman immunoglobulin G on cell surfaces at the electron microscopic level: a new method, theory and application. J Histochem Cytochem 25:1187–1200

Georgi R (1969a) Feinstruktur peritrophischer Membranen von Crustaceen. Z Morphol Ökol Tiere 65:225–273

Georgi R (1969b) Bildung peritrophischer Membranen von Decapoden. Z Zellforsch Mikrosk Anat 99:570–607

Gerber GH (1976) Reproductive behaviour and physiology of *Tenebrio molitor* (Coleoptera: Tenebrionidae). III. Histogenetic changes in the internal genitalia, mesenteron, and cuticle during sexual maturation. Can J Zool 54:990–1002

Gern L, Aeschlimann A, Zhu Z (1990) Development of *Borrelia burgdorferi* in female *Ixodes ricinus* during a blood meal. (in preparation)

Geyer G (1973) Ultrahistochemie, 2nd edn. Fischer, Stuttgart

Gibson-Carmichael T-D (1885) Notes on the anatomy of the Myriapoda. Proc R Physiol Soc Edinburgh Sess 1884–1885, pp 377–381

Giles ET (1965) The alimentary canal of *Anisolabis littorea* (White) (Dermaptera: Labiduridae), with special reference to the peritrophic membrane. Trans R Soc NZ Zool 6:87–101

Goodman SL, Hodges GM, Trejdosiewicz LK, Livingston DC (1981) Colloidal gold markers and probes for routine application in microscopy. J Microsc 123:201–213

Gordon RM (1957) *Trypanosoma congolense* in its passage through the peritrophic membrane of *Glossina morsitans*. Trans R Soc Trop Med Hyg 51:296

Gouranton J, Maillet P-L (1965) Sur l'existence d'une membrane péritrophique chez un insect suceur de sève *Cicadella viridis* L. (Homoptera, Jassidae) C R Acad Sci Paris 261:1102–1105

Gowing MM, Silver MW (1983) Origins and microenvironments of bacteria mediating fecal pellet decomposition in the sea. Mar Biol 73:7–16

Gowing MM, Silver MW (1985) Minipellets: a new and abundant size class of marine fecal pellets. J Mar Res 43:395–418

Graf R, Briegel H (1982) Comparison between aminopeptidase and trypsin activity in blood-fed females of *Aedes aegypti*. Rev Suisse Zool 89:845–850

Graham RC, Karnovsky MJ (1966) The early stages of absorption of injected horseradish peroxidase in the proximal tubules of mouse kidney: ultrastructural cytochemistry by a new technique. J Histochem Cytochem 14:291–302

Graham-Smith GS (1934) The alimentary canal of *Calliphora erythrocephala* L. with special reference to its musculature and to the proventriculus, rectal valve and rectal papillae. Parasitology 26:176–248

Grainge CA, Pearson RG (1965) Peritrophic membranes in the Phalangida. Nature (Lond) 205:925

Grandjean O (1984) Blood digestion in *Ornithodorus moubata* Murray sensu stricto Walton (Ixodoidea: Argasidae) females. I. Biochemical changes in the midgut lumen and ultrastructure of the midgut cell, related to intracellular digestion. Acarologia 25:147–165

Grell KG (1938) Der Darmtraktus von *Panorpa communis* L. und seine Anhänge bei Larve und Imago. (Ein Beitrag zur Anatomie und Histologie der Mecopteren). Zool Jahrb Anat 64:1–86

Güldner F-H (1969) Elektronenmikroskopische Untersuchungen am Intestinaltrakt von *Daphnia pulex*. Thesis, Univ Berlin

Gutiérrez LS, Burgos MH (1978) The intestine of *Triatoma infestans*. II. The surface coat of the midgut. J Ultrastruct Res 63:244–251

Hackman RH, Goldberg M (1974) Light-scattering and infra-red spectrophotometric studies of chitin and chitin derivatives. Carbohydrate Res 38:35–45

Hall CAS, Rowe G, Ryther JH, Woodwell GM (1977) Acid raid, zooplankton fecal pellets, and the global carbon budget. Biol Bull 153:427–428

Hansen MD (1978) Food and feeding behaviour of sediment feeders as exemplified by sipunculids and holothurians. Helgol Wiss Meeresunters 31:191–221

Harbison GR, McAlister VL (1979) The filter-feeding rates and particle retention efficiencies of three species of *Cyclosalpa* (Tunicata, Thaliacea). Limnol Oceanogr 24:875–892

Hardy JL, Apperson G, Asman SM, Reeves WC (1978) Selection of a strain of *Culex tarsalis* highly resistant to infection following ingestion of western equine encephalomyelytis virus. Am J Trop Med Hyg 27:313–321

Hargrave BT (1970) The effect of a deposit-feeding amphipod on the metabolism of benthic microflora. Limnol Oceanogr 15:21–30

Hargrave BT (1976) The central role of invertebrate faeces in sediment decomposition. In: Anderson JM, MacFadyen A (eds) The role of terrestrial and aquatic organisms in decomposition processes. Blackwell, Oxford, pp 301–321

218

Hargrave BT (1978) Seasonal changes in oxygen uptake by settled particulate matter and sediments in a marine bay. J Fish Res Board Can 35:1621–1628

Hargrave BT, Taguchi S (1978) Origin of deposited material sedimented in a marine bay. J Fish Res Board Can 35:1604–1613

Harley JMB (1971) Comparison of the susceptibility to infection with *Trypanosoma rhodesiense* of *Glossina pallidipes*, *G. morsitans*, *G. fuscipes*, and *G. brevipalpis*. Ann Trop Med Parasitol 65:185–189

Harmsen R (1973) The nature of the establishment barrier for *Trypanosoma brucei* in the gut of *Glossina pallidipes*. Trans R Soc Trop Med Hyg 67:364–373

Harvey ER (1980) Water and ions in the gut. In: Locke M, Smith DS (eds) Insects in the future. Academic Press, New York, pp 105–119

Harvey WR, Cioffi M, Dow JAT, Wolfersberger MG (1983) Potassium ion transport ATPase in insect epithelia. J Exp Biol 106:91–117

Haven DS, Morales-Alamo R (1967) Occurrence and transport of faecal pellets in suspension in a tidal estuary. Sediment Geol 2:141–151

Hefner RA (1929) Studies on parajulid diplopods. No. II. The micro-anatomy of the alimentary canal of *Parajulus impressus* Say. Trans Am Microsc Soc 48:321–339

Heimpel AM (1955) The pH in the midgut and blood of the larch sawfly, *Pristiphora erichsonii* (Htg.), and other insects with reference to the pathogenicity of *Bacillus cereus* Fr. and Fr. Can J Zool 33:99–106

Hennig W (1950) Grundzüge einer Theorie der phylogenetischen Systematik. Deutscher Zentralverlag, Berlin

Hennig W (1982) Phylogenetische Systematik. Parey, Hamburg

Hepburn HR (ed) (1976) The Insect Integument. Elsevier, Amsterdam, 571 pp

Hering M (1939) Die peritrophischen Hüllen der Honigbiene mit besonderer Berücksichtigung der Zeit während der Entwicklung des imaginalen Darmes. Zool Jahrb Anat 66:130–190

Heuser E, Lipp K, Wiegandt H (1974) Detection of sialic acid containing compounds and the behaviour of gangliosides in polyacrylamide-disc electrophoresis. Anal Biochem 60:382–388

Heyraud M, Fowler SW, Beasley TM, Cherry RD (1976) Polonium-210 in euphausiids: a detailed study. Mar Biol 34:127–136

Higgo JJW, Cherry RD, Heyraud M, Fowler SW (1977) Rapid removal of plutonium from the oceanic surface layer by zooplankton fecal pellets. Nature (Lond) 266:623–624

Hoare CA (1931) The peritrophic membrane of *Glossina* and its bearing on the life cycle of *Trypanosoma grayi*. Trans R Soc Trop Med Hyg 25:57–64

Holland ND, Ghiselin MT (1970) A comparative study of gut mucous cells in thirty-seven species of the class Echinoidea (Echinodemata). Biol Bull 138:286–305

Hollander F (1954) Two-component mucous barrier: its activity in protecting gastroduodenal mucosa against peptic ulceration. Arch Intern Med 93:107–129

Holliday CW, Mykles DL, Terwilliger RC, Dangott LJ (1980) Fluid secretion by the midgut caeca of the crab, *Cancer magister*. Comp Biochem Physiol 67A:259–263

Honjo S (1976) Coccoliths: production, transportation and sedimentation. Mar Micropalaeontol 1:65–79

Honjo S (1978) Sedimentation of materials in the Sargasso Sea at a 5,367 m deep station. J Mar Res 36:469–492

Honjo S, Roma M (1978) Marine copepod fecal pellets: production, preservation and sedimentation. J Mar Res 36:45–57

Hopkin SP, Nott JA (1980) Studies on the digestive cycle of the shore crab (*Carcinus maenas* L.) with special reference to the B cells in the hepatopancreas. J Mar Biol Assoc UK 60:891–907

Horisberger M (1979) Evaluation of colloidal gold as a cytochemical marker for transmission and scanning electron microscopy. Biol Cell 36:253–258

Horisberger M, Rosset J (1977) Colloidal gold, a useful marker for transmission and scanning electron microscopy. J Histochem Cytochem 25:295–305

Houk EJ, Obie F, Hardy JL (1979) Peritrophic membrane formation and the midgut barrier to arboviral infection in the mosquito, *Culex tarsalis* Coquillett (Insecta, Diptera). Acta Trop 36:39–45

Hu PC, Collier AM, Baseman JB (1976) Interaction of virulent *Mycoplasma pneumoniae* with hamster tracheal organ cultures. Infect Immun 14:217–224

Huber W (1950) Recherche sur la structure submicroscopique de la membrane péritrophique de l'intestin moyen chez quelques insectes. Arch Anat Hist Embr 33:1–20

Huber W (1954) Die submikroskopische Struktur der peritrophischen Membran aus dem Mitteldarm von *Geotrupes silvaticus* (Col.). Mitt Schweiz Entomol Ges 27:277–279

Huber W, Haasser (1950) Electron-microscope study of the peritrophic membrane in *Dixippus morosus*. Nature (Lond) 165:394

Hughes TE (1950) The physiology of the alimentary canal of *Tyroglyphus farinae*. Q J Microsc Sci 91:45–61

Hunt J, Charnley AK (1981) Abundance and distribution of the gut flora of the desert locust *Schistocerca gregaria*. J Invertebrate Pathol 38:378–385

Hunt S (1970) Polysaccharide-protein-complexes in invertebrates. Academic Press, New York

Hyman LH (1958) The occurrence of chitin in the lophophorate phyla. Biol Bull 114:106–112

Imms AD (1907) On the larval and pupal stages of *Anopheles maculipennis* Meigen. J Hyg 7:291–318

Iseki K (1981) Particulate organic matter transport to the deep sea by salp fecal pellets. Mar Ecol Prog Ser 5:55–60

Ito S (1965) The enteric surface coat on cat intestinal microvilli. J Cell Biol 27:475–491

Izuka T (1973) The pathogenic mechanism of the disease caused by *Streptococcus faecalis* AD-4 in silkworm larvae reared on the artificial diet. J Sericult Sci Jpn 41:333–337

Janse CJ, Rouwenhorst RJ, Klooster PFJ van der, Kaay HJ van der, Overdulve JP (1985) Development of *Plasmodium berghei ookinetes in the midgut of Anopheles atroparvus* mosquitoes and in vitro. Parasitology 91:219–225

Jeanne RL (1980) Evolution of social behaviour in the Vespidae. Ann Rev Entomol 25:371–396

Jeuniaux C (1963) Chitine et Chitinolyse. Masson, Paris

Johannes RE, Satomi M (1966) Composition and nutritive value of fecal pellets of a marine crustacean. Limnol Oceanogr 11:191–197

Jones JC (1960) The anatomy of rhythmical activities of the alimentary canal of *Anopheles* larvae. Ann Entomol Soc Am 53:459–474

Jones JC, Zeve VH (1968) The fine structure of the gastric caeca of *Aedes aegypti* larvae. J Insect Physiol 14:1567–1575

Jones GW, Freter R (1976) Adhesive properties of *Vibrio cholerae*: nature of the interaction with isolated rabbit brush border membranes and human erythrocytes. Infect Immunol 14:240–245

Kästner A (1933) Verdauungs- und Atemorgane der Weberknechte *Opilio parietinus* de Geer und *Phalangium opilio* L. Z Morphol Ökol Tiere 27:587–623

Keckes S, Fowler SW, Small LF (1972) Flux of different forms of [106]Ru through a marine zooplankter. Mar Biol 13:94–99

Kenchington W (1976) Adaptation of insect peritrophic membranes to form cocoon fabrics. In: Hepburn HR (ed) The insect integument. Elsevier, Amsterdam, pp 497–513

Killick-Kendrick R, Molyneux DH, Ashford RW (1974) *Leishmania* in phlebotomid sandflies. I. Modifications of the flagellum associated with attachment to the midgut and oesophageal valve of the sandfly. Proc R Soc London Ser B 187:409–419

King DG (1988) Cellular organization and peritrophic membrane formation in the cardia (proventriculus) of *Drosophila melanogaster*. J Morphol 196:253–282

Kirschner LB, Wagner S (1965) The site and permeability of the filtration locus in the crayfish antennal gland. J Exp Biol 43:385–395

Knauer GA, Martin JH, Bruland KW (1979) Fluxes of particulate carbon, nitrogen and phosphorus in the upper water column of the northeast Pacific. Deep Sea Res 26:97–108

Koch J (1938) The absorption of chloride ions by the anal papillae of Diptera larvae. J Exp Biol 15:152–160

Kofoed LH (1975) The feeding biology of *Hydrobia ventrosa* (Montague). I. The assimilation of different components of food. J Exp Mar Biol Ecol 19:1–9

Kojima J (1983) Peritrophic sac extraction in *Rhopalidia fasciata* (Hymenoptera, Vespidae). Kontyu (Tokyo) 51:502–508

Komar PD, Morse AP, Small LF, Fowler SW (1981) An analysis of sinking rates of natural copepod and euphausiid fecal pellets. Limnol Oceanogr 26:173–181

Kornicker LS (1962) Evolutionary trends among mollusk fecal pellets. J Palaeontol 36:829–834

Kornicker LS, Purdy EG (1957) A Bahamian faecal-pellet sediment. J Sediment Petrol 27:126–128

Krause M (1981) Vertical distribution of fecal pellets during FLEX 76. Helgol Meeresunters 34:313–327

Krishnan G, Ramachandran GN, Santanam MS (1955) Occurrence of chitin in the epicuticle of an arachnid, *Palamaeus swammerdami*. Nature (Lond) 176:557–558

Kümmel G (1956) Elektronenmikroskopische Untersuchungen über die chitinösen Auskleidungen der verschiedenen Abschnitte des Insektendarmes. Z Morphol Ökol Tiere 45:309–342

Kusmenko S (1940) Über die postembryonale Entwicklung des Darmes der Honigbiene und die Herkunft der larvalen peritrophischen Hüllen. Zool Jahrb Anat 66:463–530

Kusnetsova LA (1968) The cytochemistry of the mid-intestine of the female mosquitoes (*Aedes aegypti* L.) at different stages of the gonotrophic cycle. Vestn Mosk Univ Ser VI No 2:114–116

Lampitt RS, Noji T, von Bodungen B (1990) What happens to zooplankton faecal pellets? Implications for material flux. Mar Biol 104:15–23

Landis EM, Pappenheimer JR (1963) Exchange of substances through the capillary walls. In: Hamilton WF, Dow P (eds) Handbook of physiology. Circulation, vol 2. Am Physiol Soc, Washington DC, pp 961–1034

Landry MR (1981) Switching between herbivory and carnivory by the plankton marine copepod *Calanus pacificus*. Mar Biol 65:77–82

Lane NJ, Harrison JB (1979) An unusual cell surface modification: a double plasma membrane. J Cell Sci 39:355–372

Larsson R (1973) The alimentary canal of *Lepisma saccharina* L. (Thysanura, Lepismatidae). Entomol Scand 4:115–122

Lautenschlager KP, Kaushik NK, Robinson JB (1978) The peritrophic membrane and faecal pellets of *Gammarus lacustris limnaeus* Smith. Freshwater Biol 8:207–211

Lavoipierre MJ (1958) Studies on the host-parasite relationship of filarial nematodes and their arthropod hosts. II. The arthropode as a host to the nematode: a brief appraisal on our present knowledge, based on a study of the more important literature from 1878 to 1957. Ann Trop Med Parasit 52:326–345

Lee RF (1968) The histology and histochemistry of the anterior midgut of *Periplaneta americana* (Dictyoptera: Blattidae) with reference to the formation of the peritrophic membrane. Proc R Entomol Soc Lond Ser A 43 (7–9):122–134

Leger N, Cavier R (1970) A propos de l'évolution d'*Hymenolepis nana* var. *fraterna* chez des hôtes intermédiaires inhabituels. Ann Parasit (Paris) 45:195–201

Lehane MJ (1976) Formation and histochemical structure of the peritrophic membrane in the stablefly, *Stomoxys calcitrans*. J Insect Physiol 22:1551–1557

Leloup AM (1964) Cultures organotypiques de gonades d'insecte (*Calliphora erythrocephala*). Bull Soc Zool Fr 89:70–77

Levinton JS (1972) Stability and trophic structure in deposit feeding and suspension-feeding communities. Am Nat 106:472–486

Lewis DJ (1950) A peritrophic membrane in *Simulium*. Nature (Lond) 165:978

Lewis DJ (1953) *Simulium damnosum* and its relation to onchocerciasis in the Anglo-Egyptian Sudan. Bull Entomol Res 43:597–644

Loele K (1914) Beiträge zur Kenntnis der Histologie und Funktion des Hymenopterendarmes. Z Allg Physiol 16:1–36

Lopez GR, Levinton JS, Slobotkin LB (1977) The effect of grazing by the detritovore *Orchestia grillus* on *Spartina* litter and its associated microbial community. Oecologia 30:11–127

Lovett DL, Felder DL (1986) Ontogeny of gut function in *Penaeus setiferus*. Am Zool 26:34A (Abstr)

Lowry OH, Rosebrough NJ, Farr AL, Randall RJ (1951) Protein measurement with the Folin phenol reagent. J Biol Chem 193:265–275

Lozinski P (1921) Histologische Untersuchungen über den Darm der Myrmeleonidenlarve. Bull Acad Cracovie 192–222

Lyonet P (1762) Traité anatomique de la chenille. Grosse Pinet, La Haye

Madin LP (1982) Production, composition and sedimentation of salp fecal pellets in oceanic waters. Mar Biol 67:39–45

Maeda H, Ishida N (1967) Specificity of binding of hexopyranosyl polysaccharides with fluorescent brightener. J Biochem (Tokyo) 62:276–278

Maier WA (1976) Arthropoden als Wirte und Überträger menschlicher Parasiten: Pathologie und Abwehrreaktion der Wirte. Z Parasitenk 48:151–179

Maier WA, Becker-Feldmann H, Seitz HM (1987) Pathology of malaria infected mosquitoes. Parasitol Today 3:216–218

Manton SM, Heatley NG (1937) Studies on the Onychophora. II. The feeding, digestion, excretion and food storage of Peripatopsis. Philos Trans R Soc London Ser B 227:411–464

Marcus BA (1930) Untersuchungen über die Malpighischen Gefäße bei Käfern. Z Morphol Ökol Tiere 19:609–677

Marks EP, Leighton T (1980) Modes of action of inhibitors of cuticle formation in insect organ cultures. In: 16th Int Congr Entomology, Kyoto, Abstr

Marks EP, Sowa BA (1976) Cuticle formation in vitro. In: Hepburn HR (ed) The Insect integument. Elsevier, Amsterdam, pp 339–357

Marshall AT, Cheung WWK (1970) Ultrastructure and cytochemistry of an extensive plexiform surface coat of the midgut cells of a fulgorid insect. J Ultrastruct Res 33:161–172

Marshall SM, Orr AP (1972) The biology of a marine copepod. Springer, Berlin Heidelberg New York, 195 pp

Martens P (1976) Artspezifische Merkmale der Faeces von vier dominierenden Copepodenarten der Kieler Bucht. Helgol Wiss Meeresunters 28:411–416

Martignoni ME (1952) Die submikroskopische Textur der peritrophischen Membran von Peridroma margaritosa (Haw.) Mitt Schweiz Entomol Ges 25:107–110

Martin AL (1964) The alimentary canal of Marinogammarus. Proc Zool Soc (Lond) 143:525–544

Mason B, Gilbert O (1954) Presence of a peritrophic membrane in Diplopoda. Nature (Lond) 174:1022

Maudlin I (1985) Inheritance of susceptibility to trypanosomes in tsetse flies. Parasitol Today 1:59–60

McCloskey LR (1970) A new species of Dulichia (Amphipoda, Podoceridae) commensal with a sea urchin. Pac Sci 24:90–98

Mehlhorn H, Peters W, Haberkorn A (1980) The formation of kinetes and oocysts in Plasmodium gallinaceum (Haemosporidia) and considerations on phylogenetic relationships between Haemosporidia, Piroplasmida and other Coccidia. Protistologica 16:135–154

Meis JFGM, Ponnudurai T (1987) Ultrastructural studies on the interaction of Plasmodium falciparum ookinetes. Parasitol Res 73:500–506

Meis JFGM, Pool G, van Gemert GJ, Lensen AHW, Ponnudurai T, Meuwissen JHET (1989) Plasmodium falciparum ookinetes migrate intercellularly through Anopheles stephensi midgut epithelium. Parasitol Res 76:13–19

Menzi JJ (1919) Das Stomodaeum der Lumbriciden. Rev Suisse Zool 27:405–476

Mercer EH, Day MF (1952) The fine structure of the peritrophic membrane of certain insects. Biol Bull 103:384–394

Mets R de (1962) Submicroscopic structure of the peritrophic membrane in arthropods. Nature (Lond) 196:77–78

Mets R de, Jeuniaux C (1962) Sur les substances organiques constituant la membrane péritrophique des insectes. Arch Int Physiol Biochim 70:93–96

Mets R de, Nayak NA, Grégoire C (1964) On submicroscopic structure of the peritrophic membrane and some excreta of Peripatus trinidadensis (Onychophora). Proc Natl Inst Sci India B 30:131–135

Miley HH (1930) Internal anatomy of *Euryurus erythropygus* (Brandt) (Diplopoda). Ohio J Sci 30:229–249

Miller TM (1975) Neurosecretion and the control of visceral organs in insects. Ann Rev Entomol 20:133–149

Milne-Edwards H (1857–1865) Leçons sur la physiologie et l'anatomie comparée de l'homme et des animaux, 8 vols. Paris

Möbuβ A (1897) Über den Darmkanal der *Anthrenus*-Larve nebst Bemerkungen zur Epithelregeneration. Arch Naturgesch 63:89–128

Modespacher U-P, Rudin W, Jenni L, Hecker H (1986) Transport of peroxidase through the midgut epithelium of *Glossina m. morsitans* (Diptera, Glossinidae). Tissue Cell 18:429–436

Moeremans M, Daneels G, van Digek A, Langanger G (1984) Sensitive vizualization of antigen antibody reaction in dot and blot immuno gold/silver staining. J Immunol Meth 74:353–360

Moloo SK, Steiger RF, Hecker H (1970) Ultrastructure of the peritrophic membrane formation in Glossina Wiedemann. Acta Trop 27:378–394

Molyneux DH (1975) *Trypanosoma (Megatrypanum) melophagium*: modes of attachment of parasites to mid-gut, hindgut and rectum of the sheep ked, *Melophagus ovinus*. Acta Trop 32:65–74

Molynex DH, Selkirk M, Lavin D (1978) *Trypanosoma (Megatrypanum) melophagium* in the sheep ked, *Melophagus ovinus*. A scanning electron microscope (SEM) study of the parasites and the insect gut wall surfaces. Acta Trop 35:319–328

Moore HB (1931a) The muds of the Clyde Sea area. III. Chemical and physical conditions, rate and nature of sedimentation, and fauna. J Mar Biol Assoc NS 17:325–358

Moore HB (1931b) The species identification of faecal pellets. J Mar Biol Assoc NS 17:359–365

Moore HB (1931c) The systematic value of a study of molluscan faeces. Proc Malac Soc 19:281–290

Moore HB (1932) The faecal pellets of the Anomura. Proc Roy Soc Edinburgh 52:296–308

Mshelbwala AS (1972) *Trypanosoma brucei* infection in the haemocoel of the tsetse flies. Trans R Soc Trop Med Hyg 66:637–643

Munk R (1967a) Licht- und elektronenmikroskopische Befunde an der Filterkammer der Kleinzikade *Euscelis variegatus* Kbm. (Jassidae). Verh Dtsch Zool Ges 1967 Heidelberg. Zool Anz Suppl 31:519–527 (1968)

Munk R (1967b) Zur Morphologie und Histologie des Verdauungstraktes zweier Jassiden (Homoptera Auchenorrhyncha) unter besonderer Berücksichtigung der sogenannten Filterkammer. Z Wiss Zool 175:403–424

Muzzarelli RAA (1977) Chitin. Pergamon Press, Oxford, 309 pp

Myers P, Youston AA (1978) Toxic activity of *Bacillus sphaericus* SSII-1 for mosquito larvae. Infect Immunol 19:1047–1053

Nagel G, Peters W (1991) Formation, properties and degradation of the peritrophic membranes of larval and adult fleshflies, *Sarcophaga barbata* (Insecta, Diptera). Zoomorphology (in press)

Nelson JA (1924) Morphology of the honeybee larva. J Agric Res 28:1167–1213

Neville AC (1975) Biology of the arthropod cuticle. Springer, Berlin Heidelberg New York, 448 pp

Newell JE, Baxter EW (1937) On the nature of the free cell border of certain mid-gut epithelia. Q J Microsc Sci 79:123–150

Newell R (1965) The role of detritus in the nutrition of two marine deposit feeders, the prosobranch *Hydrobia ulvae* and the bivalve *Macoma balthica*. Proc Zool Soc Lond 144:25–45

Nisizawa K, Yamaguchi T, Handa N, Maeda M, Yamazaki H (1963) Chemical nature of a uronic acid-containing polysaccharide in the peritrophic membrane of the silkworm. J Biochem (Tokyo) 54:419–426

Nogge G, Gianetti M (1979) Midgut absorption of undigested albumin and other proteins by tsetse, *Glossina morsitans morsitans* (Diptera: Glossinidae). J Med Entomol 16:263

Nolan RA, Clovis CJ (1985) *Nosema fumiferanae* release into the gut of the larvae of the eastern spruce budworm *(Choristoneura fumiferana)*. J Invertebrate Pathol 45:112–114

Nopanitaya W, Misch DW (1974) Developmental cytology of the midgut in the flesh-fly *Sarcophaga bullata* (Parker). Tissue Cell 6:487–502

Norman JO, Thompson JM, Spates GE, Meola SM (1983) Microbial flora of the midgut of the stable fly, *Stomoxys calcitrans* (L.) maintained on a diet of sterile bovine blood. J Cell Biol 97:1748 (Abstr)

Nunez FS, Crawford CS (1977) Anatomy and histology of the alimentary tract of the desert millipede *Orthoporus ornatus* (Girard) (Diplopoda: Spirostreptidae). J Morphol 151:121–130

Odier A (1823) Mémoire sur la composition chimique des parties cornées des insectes. Mem Soc Hist Nat Paris 1:29–42

Oertel E (1930) Metamorphosis in the honeybee. J Morphol Physiol 50:295–339

Offek I, Mirelman D, Sharon N (1977) Adherence of *Escherichia coli* to human mucosal cells mediated by mannose receptors. Nature (Lond) 265:623–625

Omar MS (1976) Histopathologie und Abwehrmechanismen bei Simulien nach starker *Onchocerca volvulus*-Infektion. Z Angew Entomol 82:53–57

Omar MS, Garms R (1975) The fate and migration of microfilariae of a Guatemalan strain of *Onchocerca volvulus* in *Simulium ochraceum* and *S. metallicum*, and the role of the buccopharyngeal armature in the destruction of microfilariae. Tropenmed Parasit 26:183–190

Omar MS, Garms R (1977) Lethal damage to *Simulium metallicum* following high intakes of *Onchocerca volvulus* microfilariae in Guatemala. Tropenmed Parasit 28:109–119

Ono M, Kato S (1968a) Amino acid composition of the peritrophic membrane of the silkworm, *Bombyx mori* L. Bull Sericult Exp Stn Jpn 23:1–8

Ono M, Kato S (1968b) Studies on the dissolution of peritrophic membrane in the silkworm, *Bombyx mori* L. II. On the bacterial enzyme, which decomposes peritrophic membrane, obtained from the culture filtrates of *Aeromonas*. Bull Sericult Exp Sta Jpn 23:28–34

Orihel TC (1975) The peritrophic membrane: its role as a barrier to infection of the arthropod host. In: Maramorosch K, Shope RE (eds) Invertebrate immunity: mechanisms of invertebrate vector-parasite relations. Academic Press, new York, pp 65–73

Osterberg C, Carey AG, Curl H (1963) Acceleration of sinking rates of radionuclides in the ocean. Nature (Lond) 200:1276–1277

Otieno LH, Darji N, Onyango P (1976) Development of *Trypanosoma (Trypanozoon) brucei* in *Glossina morsitans*. Acta Trop 33:143–150

Owen G (1966) Digestion. In: Wilbur KM, Yonge CM (eds) Physiology of Mollusca, vol 2. Academic Press, New York, pp 53–96

Pabst MA, Crailsheim KM, Moritz B (1988) Age-dependent changes in the peritrophic membranes of the honeybee (*Apis mellifera* Linnaeus 1758): a histochemical study. Entomol Gen 14:1–10

Paffenhöfer G-A, Knowles SC (1979) Ecological implications of fecal pellet size, production and consumption by copepods. J Mar Res 37:35–49

Parsons MC (1957) The presence of a peritrophic membrane in some aquatic Hemiptera. Psyche 64:117–122

Paschke JD, Summers MD (1975) Early events in the infection of the arthropod gut by pathogenic insect viruses. In: Maramorosch K, Shope RE (eds) Invertebrate immunity: mechanisms of invertebrate vector-parasite relations. Academic Press, New York, pp 75–112

Peduzzi P, Herndl GJ (1986) Role of bacteria in decomposition of faecal pellets egested by the epiphyte-grazing gastropod *Gibbula umbilicaris*. Mar Biol 92:417–424

Perris E (1850) Note pour servir à l'histoire des *Cionus*. Ann Soc Linn Lyon 2:291–302

Perrone JB, Spielman A (1986) Microfilarial perforation of the midgut of a mosquito. J Parasitol 72:723–727

Perrone JB, Spielman A (1988) Time and site of assembly of the peritrophic membrane of the mosquito *Aedes aegypti*. Cell Tissue Res 252:473–478

Peters W (1966) Chitin in Tunicata. Experientia 22:820

Peters W (1967a) Zur Frage des Vorkommens und der Definition peritrophischer Membranen. Verh Dtsch Zool Ges Göttingen; Zool Anz Suppl 30:142–152 (1966)

Peters W (1967b) Bildung und Struktur peritrophischer Membranen bei Phalangiiden (Opiliones, Chelicerata). Z Morphol Ökol Tiere 59:134–142

Peters W (1968) Vorkommen, Zusammensetzung und Feinstruktur peritrophischer Membranen im Tierreich. Z Morphol Ökol Tiere 62:9–57

Peters W (1969) Vergleichende Untersuchungen der Feinstruktur peritrophischer Membranen von Insekten. Z Morphol Ökol Tiere 64:21–58

Peters W (1972) Occurrence of chitin in Mollusca. Comp Biochem Physiol 41B:541–550

Peters W (1976) Investigations on the peritrophic membranes of Diptera. In: Hepburn HR (ed) The Insect Integument. Elsevier, Amsterdam, pp 515–543

Peters W (1979) The fine structure of peritrophic membranes of mosquito and blackfly larvae of the genera *Aedes*, *Anopheles*, *Culex*, and *Odagmia* (Diptera: Culicidae/Simuliidae). Entomol Gen 5:289–299

Peters W, Kalnins M (1985) Aminopeptidases as immobilized enzymes on the peritrophic membranes of insects. Entomol Gen 11:25–32

Peters W, Latka I (1986) Electron microscopic localization of chitin using colloidal gold labelled with wheat germ agglutinin. Histochemistry 84:155–160

Peters W, Walldorf V (1986a) Endodermal secretion of chitin in the "cuticle" of the earthworm gizzard. Tissue Cell 18:361–374

Peters W, Walldorf V (1986b) Der Regenwurm *Lumbricus terrestris* L. Quelle & Meyer, Heidelberg

Peters W, Wiese B (1986) Permeability of the peritrophic membranes of some Diptera to labeled dextrans. J Insect Physiol 32:43–50

Peters W, Zimmermann U, Becker B (1973) Anisotropic cross-bands in peritrophic membranes of Diptera. J Insect Physiol 19:1067–1077

Peters W, Herlet N, Thienemann H (1978) Bildung und Feinstruktur peritrophischer Membranen bei Mückenlarven aus dem Bereich der Tipulo-, Psychodo- und Bibiomorpha (Diptera: Nematocera). Entomol Gen 4:33–54

Peters W, Heitmann S, D'Haese J (1979) Formation and fine structure of peritrophic membranes in the earwig, *Forficula auricularia* (Dermaptera: Forficulidae). Entomol Gen 5:241–254

Peters W, Kolb H, Kolb-Bachofen V (1983) Evidence for a sugar receptor (lectin) in the peritrophic membrane of the blowfly larva, *Calliphora erythrocephala* Mg. (Diptera). J Insect Physiol 29:275–280

Petersen H (1912) Beiträge zur vergleichenden Physiologie der Verdauung. V. Die Verdauung der Honigbiene. Pflügers Arch 145:121–151

Pilskaln CH, Honjo S (1987) The fecal pellet fraction of biogeochemical particle fluxes to the deep sea. Global Biogeochem Cycles 1:31–38

Plateau F (1878) Recherches sur les phénomènes de la digestion et sur la structure de l'appareil digestif chez les Myriapodes de Belgique. Mem Acad R Med Belg 42:719–754

Platzer-Schultz I, Reiss F (1970) Zur Histologie der Bildungszone der peritrophischen Membran einiger Chironomidenlarven (Diptera). Arch Hydrobiol 67:396–411

Platzer-Schultz I, Welsch U (1969) Zur Entstehung und Feinstruktur der peritrophischen Membranen der Larven von *Chironomus strenzkei* Fittkau (Diptera). Z Zellforsch Mikrosk Anat 100:594–605

Platzer-Schultz I, Welsch U (1970) Apokrine Sekretion der peritrophischen Membran von *Chironomus thummi piger* Str. (Diptera). Z Zellforsch Mikrosk Anat 104:530–540

Polak JM, Varndell JM (1984) Immunolabelling for electron microscopy. Elsevier, Amsterdam

Pollock EG, Koller T, Hofstadter H (1970) Localization of acid polysaccharides in electron microscopy by coupling acriflavine with PTA. 7th Int Congr EM Grenoble 1:587–588

Pomeroy LR, Deibel D (1980) Aggregation of organic matter by pelagic tunicates. Limnol Oceanogr 24:643–652

Poulson DF (1950) Histogenesis, organogenesis, and differentiation in the embryo of *Drosophila melanogaster* Meigen. In: Demerec M (Ed) Biology of *Drosophila*. Wiley, New York

Prakasam VR, Azariah J (1975) An optimum pH for the demonstration of chitin in *Periplaneta americana* using Lugol's iodine. Acta Histochem 53:238–240

Prasse J (1967) Zur Anatomie und Histologie der Acaridae mit besonderer Berücksichtigung von *Caloglyphus berlesei* (Michael 1903) und *C. michaeli* (Oudemans 1924); I. Das Darmsystem. Wiss Z Univ Halle 5:789–812

Prell H (1925) Zur Biologie der Blattschaber (Cionini). I. Die Entstehung der larvalen Gallerthülle und des Puppenkokons. Zool Anz 62:33–48

Puchta O, Wille H (1956) Ein parasitisches Bakterium im Mitteldarmepithel von *Solenobia triquetrella* F.R. (Lepid., Psychidae). Z Parasitenkd 17:400–418

Rainbow PS, Walker G (1977) The functional morphology of the alimentary tract of barnacles (Cirripedia: Thoracica). J Exp Mar Biol Ecol 28:183–206

Rajulu GS (1971) An electron microscopic study on the ultrastructure of the peritrophic membrane of a chilopod *Ethmostigmus spinosus* together with observations on its chemical composition. Curr Sci 6:134–135

Ramdohr KA (1811) Abhandlungen über die Verdauungswerkzeuge der Insekten. Hendel, Halle

Ramsay JA (1950) Osmotic regulation in mosquito larvae. J Exp Biol 27:145–157

Ravindranath MHR, Ravindranath MH (1975) A simple procedure to detect chitin in delicate structures. Acta Histochem 53:203–205

Reeve MR (1963) The filter-feeding of *Artemia*. III. Faecal pellets and their associated membranes. J Exp Biol 40:215–221

Reeve MR, Cosper TC, Walter MA (1975) Visual observations on the process of digestion and the production of faecal pellets in the chaetognath *Sagitta hispida* Conant. J Exp Mar Biol Ecol 17:39–46

Reger JF (1971) Fine structure of the surface coat of midgut epithelial cells in the homopteran, *Phylloscelis atra* (Fulgorid). J Submicrosc Cytol 3:29–37

Reid GDF, Lehane MJ (1984) Peritrophic membrane formation in three temperate simuliids, *Simulium ornatum*, *S. equinum* and *S. lineatum*, with respect to the migration of onchocercal microfilariae. Ann Trop Med Parasitol 78:527–539

Reissig JL, Strominger JL, Leloir LF (1955) A modified colorimetric method for the estimation of N-acetylamino sugars. J Biol Chem 217:959–966

Rengel C (1903) Über den Zusammenhang von Mitteldarm und Enddarm bei den Larven der aculeaten Hymenopteren. Z Wiss Zool 75:221–232

Renkin EM, Gilmore JP (1973) Glomerular filtration. In: Orloff J, Berliner RW (eds) Handbook of Physiology. Am Physiol Soc, Washington DC, pp 185–248

Reynolds SE, Nottingham SF, Stephens AE (1985) Food and water economy and its relation to growth in fifth-instar larvae of the tobacco hornworm *Manduca sexta*. J Insect Physiol 31:119–127

Richards AG (1951) The Integument of Arthropods. Univ Minnesota Press, Minneapolis

Richards AG (1975) The ultrastructure of the midgut of hematophagous insects. Acta Trop 32:83–95

Richards AG, Richards PA (1969) Development of microfibers in the peritrophic membrane of a mosquito larva. Annu Proc Electron Microsc Soc Am 27:256–257

Richards AG, Richards PA (1971) Origin and composition of the peritrophic membrane of the mosquito, *Aedes aegypti*. J Insect Physiol 17:2253–2275

Richards AG, Richards PA (1977) The peritrophic membranes of insects. Annu Rev Entomol 22:219–240

Richardson M, Romoser WS (1972) The formation of the peritrophic membrane in adult *Aedes triseriatus* (Say) (Diptera: Culicidae). J med Entomol 9:495–500

Rinderknecht HME, Geokas R, Silver, Haverback BJ (1968) A new ultrasensitive method for determination of proteolytic activity. Clin Chim Acta 21:197–203

Rizki MTM (1956) The secretory activity of the proventriculus of *Drosophila melanogaster*. J Exp Zool 131:203–220

Rodriguez DJ, Machado-Allison CE (1977) Genetica y ecologia de un nuevo mutante (peritrofica) de *Aedes aegypti* (L.). Acta Biol Venez 9:347–375

Rohringer R, Holden DW (1985) Protein blotting: detection of proteins with colloidal gold and of glycoproteins and lectins with biotin-conjugates and enzyme probes. Anal Biochem 144:118–127

226

Romoser WS (1974) Peritrophic membranes in the midgut of pupal and pre-blood meal adult mosquitoes (Diptera: Culicidae). J Med Entomol 11:397–402

Romoser WS, Cody E (1975) The formation and fate of the peritrophic membrane in adult *Culex nigripalpus* (Diptera: Culicidae) J Med Entomol 12:371–378

Romoser WS, Rothman ME (1973) The presence of a peritrophic membrane in pupal mosquitoes (Diptera: Culicidae). J Med Entomol 10:312–314

Rosin S (1946) Über Bau und Wachstum der Grenzlamelle der Epidermis bei Amphibienlarven. Analyse einer orthogonalen Fibrillenstruktur. Rev Suisse Zool 53:133–201

Roth J (1983) Application of lectin-gold complexes for electron microscopic localization of glycoconjugates on thin sections. J Histochem Cytochem 31:987–999

Rudall KM (1963) The chitin/protein complexes of insect cuticles. Adv Insect Physiol 1:257–313

Rudall KM, Kenchington W (1971) Arthropod silks: the problem of fibrous proteins in animal tissues. Annu Rev Entomol 16:73–96

Rudall KM, Kenchington W (1973) The chitin system. Biol Rev 48:597–636

Rudin W, Hecker H (1989) Lectin binding sites in the midgut of the mosquitoes *Anopheles stephensi* Liston and *Aedes aegypti* L. (Diptera: Culicidae). Parasite Res 75:268–279

Rudzinska MA, Spielman A, Lewengrub S, Presman J, Karakashian S (1982) Penetration of the peritrophic membrane of the tick by *Babesia microti*. Cell Tissue Res 221:471–481

Rungius H (1911) Der Darmkanal (der Imago und Larve) von *Dytiscus marginalis* L. Ein Beitrag zur Morphologie des Insektenkörpers. Z Wiss Zool 98:179–287

Samson Katharina (1909) Zur Anatomie und Biologie von *Ixodes ricinus* L. Z Wiss Zool 93:185–236

Santos CD, Terra WR (1986) Distribution and characterization of oligomeric digestive enzymes from *Erinnyis ello* larvae and inferences concerning secretory mechanisms and the permeability of the peritrophic membrane. Insect Biochem 16:691–700

Santos CD, Ribeiro AF, Terra WR (1986) Differential centrifugation, calcium precipitation, and ultrasonic disruption of midgut cells of *Erinnyis ello* caterpillars. Purification of cell microvilli and interferences concerning secretory mechanisms. Can J Zool 64:

Schäfer W (1953) Zur Unterscheidung gleichförmiger Kot-Pillen meerischer Evertebraten. Senckenbergiana 34:81–93

Schildmacher H (1950) Darmkanal und Verdauung bei Stechmückenlarven. Biol Zbl 69:390–438

Schlecht F (1977) Elektronenmikroskopische Untersuchungen an peritrophischen Membranen von Phyllopoden (Crustacea). Z Naturforsch 32c:462–463

Schlecht F (1979) Elektronenoptische Untersuchungen des Darmtraktes und der peritrophischen Membran von Cladoceren und Conchostracen (Phyllopoda, Crustacea). Zoomorphologie 92:161–181

Schlein Y, Lewis CT (1976) Lesions in haematophagous flies after feeding on rabbits immunized with fly tissues. Physiol Entomol 1:55–59

Schneider A (1887) Über den Darmkanal der Insekten. Zool Beitr 2:82–94

Schoenichen W (1930) Praktikum der Insektenkunde, 3rd edn. Fischer, Jena

Schrader HJ (1971) Fecal pellets: role in sedimentation of pelagic diatoms. Science 174:55–57

Schreiber M (1956) Die Abhängigkeit der Bildung der peritrophischen Hüllen von der Art der gebotenen Nahrung. Nach Untersuchungen an den Arbeiterinnen von *Apis mellifica*. Zool Beitr NF 2:1–50

Schroeder U, Arfors K-E, Tangen O (1976) Stability of fluorescein labelled dextrans in vivo and in vitro. Microvasc Res 11:33–39

Schultz TW, Kennedy JR (1976) The fine structure of the digestive system of *Daphnia pulex* (Crustacea: Cladocera). Tissue Cell 8:479–490

Schulze P (1927) Der chitinige Gespinstfaden der Larve von *Platydema tricuspis* Motsch. (Col. Tenebr.). Z Morphol Ökol Tiere 9:333–340

Scott JE, Dorling J (1969) Periodate oxidation of acid polysaccharides. III. A PAS method for chondroitin sulphates and other glycosaminoglycuronans. Histochemie 19:295–301

Scott JE, Quintarelli G, Dellovo MC (1964) The chemical and histochemical properties of Alcian blue. I. The mechanism of the Alcian blue staining. Histochemie 4:73–85

Seki H, Tsuji T, Hattori A (1974) Effect of zooplankton grazing on the formation of the anoxic layer in Tokyo Bay. Estuar Coast Mar Sci 2:145–151

Seligman AM (1965) Histochemical demonstration of some oxidized macromolecules with thiocarbohydrazide (TCH) or thiosemicarbohydrazide (TSC) and osmium tetroxide. J Histochem Cytochem 13:629–639

Shea SM (1971) Lanthanum staining of the surface coat of cells. Its enhancement by the use of fixatives containing Alcian blue or cetylpyridinium chloride. J Cell Biol 51:611–620

Silver MW, Bruland KW (1981) Differential feeding and fecal pellet composition of salps and pteropods, and the possible origin of the deep-water flora and olive-green "cells". Mar Biol 62:263–273

Singer S (1973) Insecticidal activity of recent bacterial isolates and their toxins against mosquito larvae. Nature (Lond) 244:110–111

Singer S (1974) Entomogenous bacilli against mosquito larvae. Dev Ind Microbiol 15:187–194

Sinha RN (1958) The alimentary canal of the adult *Tribolium castaneum* Herbst (Coleoptera, Tenebrionidae). J Kansas Entomol Soc 31:118–125

Skaer RJ (1981) Cellular sieving by a natural, high-flux membrane. I. The separation of platelets from plasma. J Microsc 124:331–333

Sluiters JF, Visser PE, van der Kaay JH (1986) The establishment of *Plasmodium berghei* in mosquitoes of a refractory and a susceptible line of *Anopheles atroparvus*. Z Parasitenkd 72:313–322

Small LF, Fowler SW (1973) Turnover and vertical transport of zinc by the euphausiid *Meganyctiphanthes norvegica* in the Ligurian Sea. Mar Biol 18:284–290

Small LF, Fowler SW, Ünlü MY (1979) Sinking rates of natural copepod fecal pellets. Mar Biol 51:233–241

Smayda TJ (1969) Some measurements of the sinking rate of fecal pellets. Limnol Oceanogr 14:621–625

Smayda TJ (1970) The suspension and sinking of phytoplankton in the sea. Oceanogr Mar Biol Annu Rev 8:353–414

Smayda TJ (1971) Normal and accelerated sinking of phytoplankton in the sea. Mar Geol 11:105–122

Smayda TJ (1974) Some experiments on the sinking characteristics of two freshwater diatoms. Limnol Oceanogr 19:628–635

Smetacek V (1980) Zooplankton standing stock, copepod faecal pellets and particulate detritus in Kiel Bight. Estuar Coast Mar Sci 11:477–490

Smith DS (1968) Insect Cells, their Structure and Function. Oliver Boyd, Edinburgh

Snyder SD, Wolfe AF (1980) A histological study of the digestive system of *Artemia* with reference to production of its peritrophic membrane. Proc Pa Acad Sci 54:123–127

Soltani N (1984) Effects of ingested Diflubenzuron on the longevity and the peritrophic membrane of adult mealworms (*Tenebrio molitor* L.) Pestic Sci 15:221–225

Soutar A, Kling SA, Crill PA, Duffrin E, Bruland KW (1977) Monitoring the marine environment through sedimentation. Nature (Lond) 266:136–139

Spiro RG (1963) Glycoproteins: their biochemistry, biology and role in human disease. New Engl J Med 281:999–1001, 1043–1056

Springall DR, Hacker GW, Grimelius L, Polak JM (1984) The potential of the immunogold-silver staining method for paraffin sections. Histochemistry 81:603–608

Stahl PD, Rodman JS, Miller MJ, Schlesinger P (1978) Evidence for receptor mediated binding of glycoproteins, glycoconjugates and lysosomal glycosidases by alveolar macrophages. Proc Natl Acad Sci USA 75:1399–1403

Stamm B, D'Haese J, Peters W (1978) SDS gel electrophoresis of proteins and glycoproteins from peritrophic membranes of some Diptera. J Insect Physiol 24:1–8

Steedman HF (1950) Alcian blue 8 GS: a new stain for mucin. Q J Microsc Sci 91:477–479

Stegemann H (1963) Protein (conchagen) and chitin in the supporting tissue of the cuttlefish. Z Physiol Chem 331:269–279

Steiger RF (1973) On the ultrastructure of *Trypanosoma (Trypanozoon) brucei* in the course of its life cycle and some related aspects. Acta Trop 30:64–168

228

Steinhaus EA (1967) Insect Microbiology. Hafner, New York

Stobbart RH (1971) Factors affecting the control of body volume in the larvae of the mosquitoes *Aedes aegypti* L. and *Aedes detritus* Edw. J Exp Biol 54:67–82

Stohler H (1957) Analyse des Infektionsverlaufes von *Plasmodium gallinaceum* im Darme von *Aedes aegypti*. Acta Trop 14:302–352

Stohler H (1961) The peritrophic membrane of blood sucking Diptera in relation to their role as vectors of blood parasites. Acta Trop 18:263–266

Stoltz DB, Summers MD (1971) Pathway of infection of mosquito iridescent virus. I. Preliminary observations on the fate of ingested virus. J Virol 8:900–909

Streng R (1969) Chitinhaltiger Spinnfaden bei der Larve des Buchenspringrüßlers (*Rhynchaenus fagi* L.). Naturwissenschaften 56:333–334

Streng R (1973) Die Erzeugung eines chitinigen Kokonfadens aus peritrophischer Membran bei der Larve on *Rhynchaenus fagi* L. (Coleoptera, Curculionidae). Z Morphol Ökol Tiere 75:137–164

Sutherland DR, Christensen BM, Lasee BA (1986) Midgut barrier as a possible factor in filarial worm vector competency in *Aedes trivittatus*. J Invertebrate Pathol 47:1–7

Sutter A (1977) Ultrastrukturelle Untersuchung des Mitteldarmepithels bei *Aedes aegypti*-Larven. Diplomarb, Schweiz Tropeninst, Basel

Sutton MF (1951) On the food, feeding mechanism and alimentary canal of Corixidae (Hemiptera, Heteroptera). Proc Zool Soc Lond 121:465–499

Sutton MF (1957) The feeding mechanism, functional morphology and histology of the alimentary canal of *Terebella lapidaria* L. (Polychaeta). Proc Zool Soc Lond 129:487

Swanson J, Sparks E, Young D, King G (1975) Studies on gonococcus infection. X. Pili and leukocyte association factor as mediators of interactions between gonococci and eukaryotic cells in vitro. Infect Immunol 11:1352–1361

Taghon GL, Nowell ARM, Jumars PA (1984) Transport and breakdown of fecal pellets: biological and sedimentological consequences. Limnol Oceanogr 29:64–72

Tanada Y, Hess RT, Omi EM (1975) Invasion of a nuclear polyhedrosis virus in midgut of the armyworm, *Pseudaletia unipuncta*, and the enhancement of a synergistic enzyme. J Invertebrate Pathol 26:99–104

Terra WR, Ferreira C (1981) The physiological role of the peritrophic membrane and trehalase: digestive enzymes in the midgut and excreta of starved larvae of *Rhynchosciara*. J Insect Physiol 27:325–331

Terra WR, Ferreira C (1983) Further evidence that enzymes involved in the final stages of digestion by *Rhynchosciara* do not enter the endoperitrophic space. Insect Biochem 13:143–150

Terra WR, Ferreira C, De Bianchi AG (1979) Distribution of digestive enzymes among the endo- and ectoperitrophic spaces and midgut cells of *Rhynchosciara* and its physiological significance. J Insect Physiol 25:487–494

Terra WR, Ferreira C, Bastos F (1985) Phylogenetic considerations of insect digestion disaccharidases and the spatial organization of digestion in the *Tenebrio molitor* larvae. Insect Biochem 15:443–450

Thiéry JP (1967) Mise en évidence des polysacharides sur coupes fines en microscopie électronique. J Microsc (Paris) 6:987–1018

Tieszen K, Molyneux DH, Abdel-Hafez SK (1985) Ultrastructure of cyst formation in *Blastocrithidia familiaris* in *Lygaeus pandurus* (Hemiptera: Lygaeidae). Z Parasitenkd 71:179–188

Tieszen K, Molyneux DH, Abdel-Hafez SK (1986) Host-parasite relationships of *Blastochrithidia familiaris* in *Lygaeus pandurus* Scop. (Hemiptera: Lygaeidae). Parasitology 92:1–12

Towbin H, Staehelin T, Gordon J (1979) Electrophoretic transfer of proteins from polyacrylamide gels to nitrocellulose sheets: procedure and some applications. Proc Natl Acad Sci USA 76:4350–4354

Treherne J (1958) The absorption and metabolism of some sugars in the locust, *Schistocerca gregaria* Forsk. J Exp Biol 35:611–625

Tristram JN (1977) Normal and cocoon-forming peritrophic membrane in larvae of the beetle *Gibbium psylloides*. J Insect Physiol 23:79–87

Tristram JN (1978) The peritrophic membrane and cocoon ribbons in larvae of *Cionus scrophulariae*. J Insect Physiol 24:391–398

Tulk A (1843) Upon the anatomy of *Phalangium opilio*. Annu Mag Nat Hist 12:243–253

Turner JT (1977) Sinking rates of fecal pellets from the marine copepod *Pontella meadii*. Mar Biol 40:249–259

Turner JT (1979) Microbial attachment to copepod fecal pellets and its possible ecological significance. Trans Am Microsc Soc 98:131–135

Turner JT (1984) Zooplankton feeding ecology: contents of fecal pellets of the copepods *Eucalanus pileatus* and *Paracalanus quasimodo* from continental shelf waters of the Gulf of Mexico. Mar Ecol Prog Ser 15:27–46

Turner JT (1986) Zooplankton feeding ecology: content of fecal pellets of the copepod *Undinula vulgaris* from continental shelf and slope waters of the Gulf of Mexico. PSZN Mar Ecol 7:1–14

Turner JT, Ferrante JG (1979) Zooplankton fecal pellets in aquatic ecosystems. Bioscience 29:670–677

Ulrich RG, Buthala DA, Klug MJ (1981) Microbiota associated with the gastrointestinal tract of the common house cricket, *Acheta domestica*. Appl Environ Microbiol 41:246–254

Vernberg FJ, Vernberg WB (eds) Pollution and physiology of marine organisms. Academic Press, New York

Vernon GM, Herold L, Witkus ER (1974) Fine structure of the digestive tract epithelium in the terrestrial isopod, *Armadillidium vulgare*. J Morphol 144:337–360

Vierhaus H (1971) Über peritrophische Membranen und andere chitinhaltige Strukturen bei Anneliden, unter besonderer Berücksichtigung von *Lumbricus terrestris* L. Thesis, Univ Düsseldorf.

Vijayambika V, John PA (1975) Functional morphology of the alimentary canal of the fish mite *Lardoglyphus konoi*. Forma Funct 8:387–394

Volkmann A, Peters W (1989a) Investigations on the midgut caeca of mosquito larvae. I. Fine structure. Tissue Cell 21:243–251

Volkmann A, Peters W (1989b) Investigations on the midgut caeca of mosquito larvae. II. Functional aspects. Tissue Cell 21:253–261

Wainwright SA, Biggs WD, Currey JD, Gosline JM (1976) Mechanical Design in Organisms. Arnold, London, 423 pp

Walker VK, Geer W, Williamson JH (1980) Dietary modulation and histochemical localization of leucine aminopeptidase activity in *Drosophila melanogaster*. Insect Biochem 10:543–548

Waterhouse DF (1952) Studies on the digestion of wool by insects. VII. Some features of digestion in three species of dermestid larvae and a comparison with *Tineola* larvae. Aust J Sci Res B 5:444–459

Waterhouse DF (1953a) The occurrence and significance of the peritrophic membrane, with special reference to adult Lepidoptera and Diptera. Aust J Zool 1:299–318

Waterhouse DF (1953b) Occurrence and endodermal origin of the peritrophic membrane in some insects. Nature (Lond) 172:676

Waterhouse DF (1954) The rate of production of the peritrophic membrane in some insects. Aust J Biol Sci 7:59–72

Weil E (1936) Vergleichend-morphologische Untersuchungen am Darmkanal einiger Apiden und Vespiden. Z Morphol Ökol Tiere 30:438–478

Wharton GW, Brody AR (1972) The peritrophic membrane of the mite, *Dermatophagoides farinae*: Acariformes. J Parasitol 58:801–804

Whitfield SG, Murphy FA, Sudia WD (1973) St. Louis encephalitis virus: an ultrastructural study of infection in a mosquito vector. Virology 56:70–87

Wiebe PH, Boyd SH, Winget C (1976) Particulate matter sinking to the deep-sea floor at 2000 m in the tongue of the ocean, with description of a new sediment trap. J Mar Res 34:341–354

Wiebe PH, Madin LP, Haury LR, Harbison GR, Philbin LM (1979) Diel vertical migration by *Salpa apera* and its potential for large-scale particulate organic matter transport to the deep-sea. Mar Biol 53:249–255

Wieser W (1965) Untersuchungen über die Ernährung und den Gesamtstoffwechsel von *Porcellio scaber* (Crustacea: Isopoda). Pedoibiologie 5:304–331

Wieser W (1966) Copper and the role of isopods in degradation of organic matter. Science 153:67–69

Wigglesworth VB (1929) Digestion in the tsetse-fly: a study of structure and function. Parasitology 21:288–321

Wigglesworth VB (1930) The formation of the peritrophic membrane in insects, with special reference to the larvae of mosquitoes. Q J Microsc Sci 73:593–616

Wigglesworth VB (1933a) The function of the anal gills of mosquito larvae. J Exp Biol 10:16–26

Wigglesworth BV (1933b) The adaptation of mosquito larvae to salt water. J Exp Biol 10:27–37

Wigglesworth VB (1942) The storage of protein, fat, glycogen and uric acid in the fat body and other tissues of mosquito larvae. J Exp Biol 19:56–77

Wigglesworth BV (1972) The Principles of Insect Physiology, 7th edn. Chapman Hall, London

Wildbolz T (1954) Beitrag zur Anatomie, Histologie und Physiologie des Darmkanals der Larve von *Melolontha melolontha* L. Mitt Schweiz Entomol Ges 27:193–240

Willett KC (1966) Development of the peritrophic membrane in *Glossina* (tsetse flies) and its relation to infection with trypanosomes. Exp Parasitol 18:290–295

Wirén A (1885) Om circulations-och digestions-organen hos Annelider af Familjerna Ampharetidae, Terebellidae, och Amphictenidae. Svensk Vet Agad Handlg NF 21

Wisselingh G van (1898) Mikrochemische Untersuchungen über die Zellwände der Fungi. Jahrb Wiss Bot 31:619–687

Wolfersberger MG, Spaeth DD, Dow JAT (1986) Permeability of the peritrophic membrane of tobacco hornworm larval midgut. Am Zool 26:74A (Abstr)

Wyatt GR (1956) Culture in vitro of tissue from the silkworm *Bombyx mori*. J Gen Physiol 39:841–852

Yaguzinskaya LW (1940) Presence of a peritrophic membrane in the midgut of the female *Anopheles maculipennis*. Med Paras (Moscow) 9:601–602 (in Russian)

Yamamoto H, Ogura N, Chigusa Y (1983) Studies on filariasis. II. Exsheathment of the microfilariae of *Brugia pahangi* in *Armigeres subalbatus*. Jpn J Parasitol 32:287–292

Yang Y, Davies DM (1977) The peritrophic membrane in adult simuliids (Diptera) before and after feeding on blood and blood-sucrose mixtures. Entomol Exp Appl 22:132–140

Yeates RA, Steiger S (1981) Ultrastructural damage of in vitro cultured ookinetes of *Plasmodium gallinaceum* (Brumpt) by purified proteinases of susceptible *Aedes aegypti* (L.). Z Parasitenkd 66:93–98

Youngbluth MJ (1982) Utilization of a faecal mass as food by the pelagic mysis larva of the penaeid shrimp *Solenocera atlantidis*. Mar Biol 66:47–51

Yunovitz H, Sneh B, Schuster S, Oron U, Broza M, Yawetz A (1986) A new sensitive method for determining the toxicity of a highly purified fraction from δ-endotoxin produced by *Bacillus thuringiensis* var. entomocidus on isolated larval midgut of *Spodoptera littoralis* (Lepidoptera, Noctuidae). J Invertebrate Pathol 48:223–231

Zachary A, Colwell RR (1979) Gut-associated microflora of *Limnoria tripunctata* in marine creosote-treated wood pilings. Nature (Lond) 282:716–717

Zachary A, Parrish KK, Bultman JD (1983) Possible role of marine bacteria in providing the creosote-resistance of *Limnoria tripunctata*. Mar Biol 75:1–8

Zachvatkin AA (1959) Fauna of USSR, Arachnoidea. 473 pp (Transl from Russian for AIBS, Washington DC, by Ratcliffe A, Hughes AM)

Zhuzhikov DP (1962) The formation of peritrophic membranes in mosquitoes *Aedes aegypti* L. Nauch Dokl Vyssch Shk Biol 4:25–27 (in Russian)

Zhuzhikov DP (1963) The structure of peritrophic membranes of Diptera. Vestn Mosk Univ 1963:24–35 (in Russian)

Zhuzhikov DP (1964) Function of the peritrophic membrane in *Musca domestica* L. and *Calliphora erythrocephala* Meig. J Insect Physiol 10:273–278

Zhuzhikov DP (1966) Investigation of the peritrophic membrane of some Diptera in polarized light. Vestn Mosk Univ 1966:37–41 (in Russian)

Zhuzhikov DP (1970) Permeability of the peritrophic membrane in the larvae of *Aedes aegypti*. J Insect Physiol 16:1193–1202

Zhuzhikov DP (1971) The structure and permeability of the peritrophic membrane in the larva of *Aedes aegypti* L. In: Proc 13th Int Congr Entomology Moscow 1968, vol 1. Nauka, Leningrad 1971, pp 463–464

Zimmermann D, Peters W (1987) Fine structure and permeability of peritrophic membranes of *Callipora erythrocephala* (Meigen) (Insecta: Diptera) after inhibition of chitin and protein synthesis. Comp Biochem Physiol 86B:353–360

Zimmermann U, Hallstein H (1970) Untersuchungen über den Transport von Stoffen durch peritrophische Membranen. II. Regelmäßige Bandenstruktur in peritrophischen Membranen. Z Naturforsch 25B:1155–1157

Zimmermann U, Mehlan D (1976) Water transport across peritrophic membranes of *Calliphora erythrocephala*. Comp Biochem Physiol 55A:119–126

Zimmermann U, Peters W, Hallstein H (1969) Untersuchungen über den Transport von Stoffen durch peritrophische Membranen. I. Struktur und Bildungsgeschwindigkeit peritrophischer Membranen von *Calliphora erythrocephala* Mg. (Diptera). Z Naturforsch 24B:1456–1460

Zimmermann U, Mehlan D, Peters W (1973) Investigations on the transport function and structure of peritrophic membranes. III. Periodic incorporation of glucose, methionine, and cysteine into the peritrophic membranes of the blowfly *Calliphora erythrocephala* Mg. in vivo and in vitro. Comp Biochem Physiol 45B:683–693

Zimmermann U, Mehlan D, Peters W (1975) Investigations on the transport function and structure of peritrophic membranes. V. Amino acid analysis and electron microscopic investigations of the peritrophic membranes of the blowfly *Calliphora erythrocephala* Mg. Comp Biochem Physiol 51B:181–186

Subject Index

236

Printing: COLOR-DRUCK DORFI GmbH, Berlin
Binding: Buchbinderei Lüderitz & Bauer, Berlin

DATE DUE

DEMCO, INC. 38-2971